Nursery Screening for *Ganoderma* Response in Oil Palm Seedlings:

A Manual

Techniques in Plantation Science Series

Series editors:

Brian P. Forster, Lead Scientist, Verdant Bioscience, Indonesia
Peter D.S. Caligari, Science Strategy Executive Director, Verdant Bioscience, Indonesia

About the series:

A series of manuals covering techniques in plantation science that form the essential underlying needs to carry out plantation science.

The series reflects the expertise in Verdant Bioscience that underlies the plantation science activities carried out at the Verdant Plantation Science Centre at Timbang Deli, Deli Serdang, North Sumatra, Indonesia.

Titles available:

1. Crossing in Oil Palm: A Manual – Umi Setiawati, Baihaqi Sitepu, Fazrin Nur, Brian P. Forster and Sylvester Dery
2. Seed Production in Oil Palm: A Manual – Eddy S. Kelanaputra, Stephen P.C. Nelson, Umi Setiawati, Baihaqi Sitepu, Fazrin Nur, Brian P. Forster and Abdul R. Purba
3. Nursery Screening for *Ganoderma* Response in Oil Palm Seedlings: A Manual – Miranti Rahmaningsih, Ike Virdiana, Syamsul Bahri, Yassier Anwar, Brian P. Forster and Frédéric Breton
4. Mutation Breeding in Oil Palm: A Manual – Fazrin Nur, Brian P. Forster, Samuel A. Osei, Samuel Amiteye, Jennifer Ciomas, Soeranto Hoeman and Ljupcho Jankuloski

Nursery Screening for *Ganoderma* Response in Oil Palm Seedlings:

A Manual

Miranti Rahmaningsih
Verdant Bioscience, Indonesia

Ike Virdiana
Verdant Bioscience, Indonesia

Syamsul Bahri
Verdant Bioscience, Indonesia

Yassier Anwar
Verdant Bioscience, Indonesia

Brian P. Forster
Verdant Bioscience, Indonesia

Frédéric Breton
CIRAD, France

CABI is a trading name of CAB International

CABI	CABI
Nosworthy Way	745 Atlantic Avenue
Wallingford	8th Floor
Oxfordshire OX10 8DE	Boston, MA 02111
UK	USA
Tel: +44 (0)1491 832111	Tel: +1 (617)682-9015
Fax: +44 (0)1491 833508	E-mail: cabi-nao@cabi.org
E-mail: info@cabi.org	
Website: www.cabi.org	

A catalogue record for this book is available from the British Library, London, UK.

Library of Congress Cataloging-in-Publication Data

Names: Rahmaningsih, Miranti, author.
Title: Nursery screening for Ganoderma response in oil palm seedlings : a manual / Miranti Rahmaningsih, Ike Virdiana, Syamsul Bahri, Yassier Anwar, Brian P. Forster, Frederic Breton.
Description: Boston, MA : CABI, [2018] | Series: Techniques in plantation science series ; 3 | Includes bibliographical references and index.
Identifiers: LCCN 2018016841 (print) | LCCN 2018023044 (ebook) | ISBN 9781786396259 (ePDF) | ISBN 9781786396266 (ePub) | ISBN 9781786396242 (pbk : alk. paper)
Subjects: LCSH: Oil palm--Diseases and pests. | Oil palm--Seedlings--Evaluation. | Ganoderma diseases of plants.
Classification: LCC SB608.O27 (ebook) | LCC SB608.O27 R34 2018 (print) | DDC 633.8/51--dc23
LC record available at https://lccn.loc.gov/2018016841

ISBN-13: 978 1 78639624 2 (pbk)

Commissioning editor: Rachael Russell
Editorial assistant: Emma McCann
Production editor: James Bishop

Typeset by SPi, Pondicherry, India
Printed and bound in the UK by Severn, Gloucester.

Series Foreword – Techniques in Plantation Science

Verdant Bioscience, Singapore (VBS), is a new company established in October 2013 with a vision to develop high-yielding, high-quality planting material in oil palm and rubber through the application of sound practices based on scientific innovation in plant breeding. The approach is to fuse traditional breeding strategies with the latest methods in biotechnology. These techniques are integrated with expertise and the application of sustainable aspects of agronomy and crop protection, alongside information and imaging technology which not only find relevance in direct aspects of plantation practice but also in selection within the breeding programme. When high-yielding planting material is allied with efficient plantation practices, it leads to what may be termed 'intensive sustainable' production. At the same time, the quality of new products is refined to give more specialized uses alongside more commodity-based oil production, thus meeting the market demands of the modern world community, but with a minimal harmful footprint. An essential ingredient in all this is having sound and practical protocols and techniques to allow the realization of the strategies that are envisaged.

To achieve its aims, VBS acquired an Indonesian company called PT Timbang Deli Indonesia, with an estate of over 970 ha of land at Timbang Deli, Deli Serdang, North Sumatra, Indonesia, and the group works under the name of 'Verdant'. A central part of this estate, which will be used for important plant nurseries and field trials, is the development of the Verdant Plantation Science Centre (VPSC), to which the operational staff moved in October 2016. A seed production and marketing facility is now established at VPSC for commercial seed sales and the processing of seed from breeding programmes. The centre comprises specialized laboratories in cell biology, genomics, tissue culture, pollen, soil DNA, plant and soil nutrition, bunch and oil, agronomy and crop protection. Field facilities include extensive nurseries, seed gardens and trials (trial sites are also located at various locations across Indonesia). It is the aim of the company to use its existing and rapidly developing intellectual property (IP) to develop superior cultivars that not only have outstanding yield but also are resistant to both biotic and abiotic stresses, while at the same time meeting new market demands. Verdant not only develops and supplies superior planting materials but also supports its customers and growers with a package of services and advice in fertilizer recommendations and crop protection. This is all part of a central mission to promote green, eco-friendly agriculture.

Brian P. Forster and Peter D.S. Caligari
Lead Scientist and Science Strategy Executive Director
Verdant Bioscience

Contents

Acknowledgements

The authors are grateful to colleagues in the Verdant Bioscience Breeding and Crop Protection Departments for sharing their knowledge and providing helpful advice in preparing this manual.

Preface

As noted in the Foreword to this Series, a central objective in Verdant Bioscience's mission is to sell better varieties of oil palm, rubber and other plantation crops through plant breeding. Essential to this objective is that the seedlings fulfil as far as possible their genetic potential, while at the same time being part of a sustainable plantation system. Therefore, one of the major breeding objectives for oil palm improvement, in line with the major need for sustainability, is the development of disease-resistant varieties. *Ganoderma* is the biggest disease problem of oil palm in South-east Asia and the number one target for disease resistance breeding in this part of the world. Interestingly, in many of the African countries, the major disease problem is *Fusarium* rather than *Ganoderma*. Effective breeding requires the identification of germplasm resistant to *Ganoderma* and the screening and selection of progenies and palms from crossing programmes in order to improve the resistance levels in the commercially grown seed. Early and rapid screens are needed to allow a sufficiently effective and high selection pressure to be imposed in order to increase the chances of identifying usable levels of resistance. Here, we present methods in a seedling screen for *Ganoderma* resistance/susceptibility in oil palm, which means large numbers of seedlings can be screened at the nursery level, and hence allow large numbers to be handled. The results of such selection will give palms, families and populations that can be used in further selection or advancement to commercialization. These then will be grown in a system that also uses appropriate husbandry techniques and biological controls, which in unison prevent or reduce *Ganoderma* incidence, in other words, in integrated management systems. Our target audiences are students and researchers in agriculture, plant pathologists, plant breeders, growers and end-users interested in the practicalities of oil palm breeding.

Brian P. Forster and Peter D.S. Caligari
Series Editors
February 2018

Introduction

1

Abstract

Oil palm is frequently planted as a continuous monoculture in countries where it is cultivated. This system brings some advantages as it allows standardized agronomic practices in optimizing crop yield per unit of land. However, there are also numerous disadvantages. One of the biggest disadvantages of monoculture is increased susceptibility to pests and diseases. Currently in South-east Asia, oil palm plantations suffer from basal stem rot (BSR) disease caused by the basidiomycete soil-borne fungus, *Ganoderma* spp. The disease is also known in other regions (West Africa and South America), although the incidences are much lower than in South-east Asia (Indonesia and Malaysia). Agronomic approaches have been developed to reduce the impact of the disease. The development of integrated pest management (IPM) practices, by combining chemical, physical and biological controls, has been pursued, but has not shown an overall significant reduction in the incidences of the disease. Breeding to develop *Ganoderma*-resistant material is considered the best additional weapon to overcome the spread of BSR disease. Field selection of mature palms is commonly used, but this can take 10–12 years of observations. Therefore, the development of a quick and early screen is of particular interest. What is described in this manual involves the artificial exposure of seedlings to the disease organism in a nursery screen and recording responses.

1.1 *Ganoderma* Disease of Oil Palm

In the past three decades, there have been three diseases particularly that have caused devastation to the oil palm industry: *Ganoderma* disease in South-east Asia, vascular wilt by *Fusarium oxysporum* f. sp. *elaeidis* in Africa and bud rot disease by *Phytophthora palmivora* in South America (Kushairi, 2012; Rajanaidu *et al.*, 2012). These three diseases have caused not only palm stand losses but also significant economic losses. *Ganoderma* disease can cause palm stand losses of up

© Miranti Rahmaningsih, Ike Virdiana, Syamsul Bahri, Yassier Anwar, Brian P. Forster and Frédéric Breton 2018. *Nursery Screening for Ganoderma Response in Oil Palm Seedlings: A Manual*

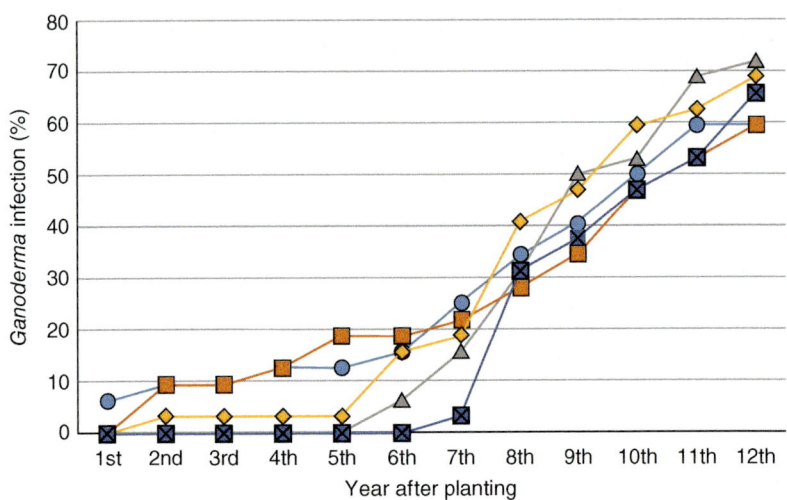

Fig. 1.1. The increased *Ganoderma* infection of some progenies in a *Ganoderma* screening trial in Indonesia. (Redrawn from Setiawati *et al.*, 2014).

to 80% (Fig. 1.1) and US$500 million in economic losses per year in Indonesia and Malaysia. In Africa, palm stand reductions of up to 50% and yield reductions of up to 30% were recorded due to vascular wilt disease 10 years after replanting (Flood, 2006; Breton *et al.*, 2010; Hushiarian *et al.*, 2013). Two outbreaks of bud rot in Colombia in 2014 caused losses of US$250 million (Torres *et al.*, 2016).

In South-east Asia, *Ganoderma* causes both basal stem rot (BSR) and upper stem rot (USR) diseases, although another fungus, *Phellinus noxius*, is suspected as the first disease agent of USR (Flood *et al.*, 2000; Hushiarian *et al.*, 2013; Rakib *et al.*, 2014). Of the two diseases, BSR causes more destruction than USR. USR is generally considered a relatively minor disease and has less frequent manifestations (Rees *et al.*, 2012; Rakib *et al.*, 2014). However, the USR incidence in some plantation estates in North Sumatra is increasing (Rees *et al.*, 2012). It is reported that the ratio of BSR to USR in plantation estates ranges from 10:1 to 1:1 (Hushiarian *et al.*, 2013).

1.2 Basal Stem Rot Disease

Although BSR is described as the main disease of oil palm in South-east Asia, the disease was first discovered in Zaire (now the Democratic Republic of the Congo) in 1915. At that time, the disease was not considered a danger to crop production (Turner, 1981). Today, BSR has become the most devastating disease of oil palm, especially in South-east Asia. Although the disease is found in every country that cultivates oil palm on a large scale, Indonesia and Malaysia are the two countries that suffer the most from BSR. Consequently, these countries are active in seeking ways to reduce *Ganoderma* disease (Cooper *et al.*, 2011; Hama-Ali *et al.*, 2014; Rakib *et al.*, 2015).

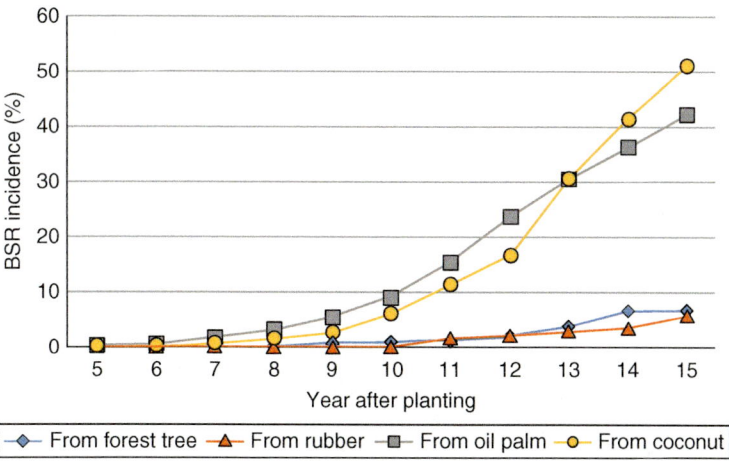

Fig. 1.2. Incidence of BSR disease in oil palm in relation to previous crops. (Redrawn from Singh, 1991).

Continuous monoculture favours disease, and the incidence of BSR has risen sharply in the past two decades, from being very low with little impact on yield to becoming a major devastating disease of oil palm. In the first cycle of oil palm plantations (in which the previous crop is not oil palm), the disease only infects older palms and therefore has no significant impact on the number of dead palms and yield loss. However, after successive replanting from one cycle of oil palm to the next cycle of oil palm (a common practice), the disease occurs increasingly early in younger palms, resulting in severe yield losses (Fig. 1.2) (Singh, 1991; Breton *et al.*, 2010; Purba *et al.*, 2012; Nuranis *et al.*, 2016).

1.3 Cause of Basal Stem Rot Disease

Basal stem rot is caused by the basidiomycete soil-borne fungus, *Ganoderma* spp. Infection is thought to be by root contact with infected palms or decaying plant debris. Spores are thought to be the main means of disease spread from plantation to plantation but, once established, one strain predominates (Breton *et al.*, 2006), suggesting spread by mycelia within a plantation. However, the recent findings by Merciere *et al.* (2017) show a high genetic diversity within some areas with a high incidence of *Ganoderma*. The fungus not only causes BSR, but also upper stem rot (USR) disease. The USR incidence supports the epidemiology theory of disease spread by airborne spores. Infected young palms usually die within 1 or 2 years after the appearance of the first symptom (basidiocarps at the base of the tree) (Fig. 1.3a), while mature trees can survive for 3–4 years (Fig. 1.3b) (Hushiarian *et al.*, 2013).

Fig. 1.3. (a) Fallen palm infected by *Ganoderma*; (b) standing palm infected by *Ganoderma*.

1.4 *Ganoderma* Diversity

Ganoderma species that cause BSR vary in different regions. Although the common species in West Africa is generally reported to be *Ganoderma lucidum* Karts, the predominant disease species in Nigeria, which is also part of West Africa, are *Ganoderma encidum*, *Ganoderma colossus* and *Ganoderma applanatum*. In South-east Asia (Indonesia and Malaysia), where the disease is the most devastating, the primary causative species is *Ganoderma boninense* (Susanto, 2002; Hushiarian *et al.*, 2013). The origin of *G. boninense* is still not known – whether the fungus came from Africa or originally was in South-east Asia (Merciere *et al.*, 2017). Moreover, studies by Wong *et al.* (2012) and Rakib *et al.* (2014) found that other *Ganoderma* species, *Ganoderma zonatum*, *Ganoderma miniatocinctum* and *Ganoderma tornatum*, were also present in diseased oil palms in Malaysia.

The first identified *Ganoderma* species was *G. lucidum* in 1881, identified by a Finnish mycologist, Peter Adolf Karsten, although its origin was not mentioned (Seo and Kirk, 2000; Susanto, 2002). The identification was based on host specificity, geographic distribution and the macromorphology of the bracket fruiting body (basidiocarp). The next identification was based on the morphology of spores, hyphal and basidiocarp characteristics. Currently, there are more than 300 *Ganoderma* species identified (Susanto, 2002). Genetic markers such as internal transcribed spacers (ITSs) of the nuclear rDNA can be used to distinguish between and within species of *Ganoderma*. Based on this method, *Ganoderma* is divided into three large groups and some subgroups (Table 1.1). *G. boninense* is in Group 2, Subgroup 2.1 (Moncalvo, 2000). However, it is interesting that *G. miniatocinctum* is not described clearly in this table, although it has been associated with disease. Another category described by Steyaert (1967) distinguishes *Ganoderma* based on the existence of laccate. *Ganoderma* with laccate (varnished or polished) can be recognized by their shiny, reddish-brown and shelf-like basidiocarp (Lyod *et al.*, 2017). *G. miniatocinctum* is placed in the laccate *Ganoderma* category (Steyaert, 1967).

A recent study based on mating compatibility reveals that *G. boninense* is 75% compatible with *G. miniatocinctum*. This percentage could represent different strains of the same species. However, there was not enough evidence to conclude that *G. boninense* and *G. miniatocinctum* represented different species or strains, but morphological studies of both fungi suggested that they might be regarded as synonymous (Jing *et al.*, 2015). The biodiversity of *Ganoderma* is now being studied and characterized at the DNA level. This allows the identification of different species and strains within a species (Zaremski *et al.*, 2013).

Table 1.1. Taxanomic groupings of *Ganoderma* based on DNA analysis of internal transcribed spacer region. (From Moncalvo, 2000.)

Group	Geographic categories	Hosts
Group 1		
1.1 *Ganoderma lucidum* complex *sensu stricto*		
– *G. lucidum*	Europe	Woody dicots, Conifers
– *Ganoderma valesiacum*	Europe	Conifers
– *Ganoderma carnosum* (= *G. atkinsonii*)	Europe	Conifers
– *Ganoderma ahmadii*	India, Pakistan	Woody dicots
– *Ganoderma tsugae*	China, Korea, North America	Conifers
– *Ganoderma oregonense*	North America	Conifers
– *Ganoderma praelongum, Ganoderma oerstedii*	South America	Woody dicots
1.2 *Ganoderma resinaceum* complex *sensu lato* *G. resinaceum* complex *sensu stricto*:		
– *G. resinaceum* ('*Ganoderma pfeifferi*')	South Africa, Europe	Woody dicots
– *G.* cf. *resinaceum* ('*Ganoderma lucidum*')	Florida, North America	Woody dicots
– *G.* cf. *resinaceum* (*Ganoderma sessile, Ganoderma platense*)	South America	Woody dicots
– *Ganoderma weberianum* complex:		
– *G. weberianum* (= *Ganoderma microsporum*)	Taiwan, South-east Asia, Australia	Woody dicots
– *Ganoderma* cf. *capense*	China, Korea, South-east Asia	Woody dicots
– *Ganoderma* sp.	South-east Asia	Woody dicots
– *Ganoderma* sp. ('*Ganoderma subamboinense*')	South America, Neotropics	Woody dicots
– *Ganoderma trengganuense*	South-east Asia	Woody dicots
1.3 *Ganoderma curtisii* complex:		
– *G. curtisii* (= *Ganoderma meredithae*)	Neotropics, Florida, North America	Woody dicots, Conifers
– *G. curtisii* (*Ganoderma fulvellum*, '*Ganoderma tsugae*')	China, Korea, Japan, Taiwan, South-east Asia	Woody dicots
1.4 *Ganoderma tropicum* complex *sensu lato*:		
– *Ganoderma* sp. 'clade A'	South-east Asia, Neotropics	Woody dicots
– *Ganoderma* sp. 'clade B' ('*Ganoderma lucidum*')	India, Pakistan, Taiwan, South-east Asia	Woody dicots

Continued

Table 1.1. Continued.

Group	Geographic categories	Hosts
– *Ganoderma* sp. 'clade C'	Indo, PNG	Woody dicots
– *Ganoderma* sp.	South Africa	Woody dicots
– *Ganoderma* sp.	Neotropics	Woody dicots
– *G. tropicum* complex *sensu stricto* (*Ganoderma fornicatum*)	Taiwan, South-east Asia	Woody dicots, palms
Group 2		
2.1 'palm clade':		
– *Ganoderma zonatum-boninense* group:		
– *G. zonatum*	Florida	Palms
– *Ganoderma* sp.	South Africa, India, Pakistan	Palms
– *Ganoderma boninense*	Japan, South-east Asia, Indo, PNG, Australia	Palms
– *Ganoderma* sp.	South-east Asia, Australia	Woody dicots, palms
2.2 *Ganoderma* species:		
– *Ganoderma* sp.	South Africa	Woody dicots
– *Ganoderma* sp.	Neotropics	Woody dicots
– *Ganoderma* sp. ('*Ganoderma* cf. *tornatum*')	South-east Asia	Woody dicots
2.3 *Ganoderma* cf. *balabacense*	South Africa, South-east Asia	Woody dicots
2.4 *Ganoderma* sp.	South-east Asia	Woody dicots
2.5 *Ganoderma sinense* (= *Ganoderma formosanum* = ? *Ganoderma neojaponicum*)	China, Korea, Taiwan	Woody dicots
Group 3		
Ganoderma austral-applanatum complex *sensu lato*:		
Ganoderma applanatum A (*Ganoderma lobatum*, *Ganoderma adspersum*):	Europe, Japan, North America	Woody dicots
– *Ganoderma cupreolaccatum* (=*Ganoderma pfeifferi*)	Europe	Woody dicots
– *Ganoderma australe* complex *sensu stricto*:		
• *Ganoderma australe* complex A:		
■'Clade A.1'	China, Korea, Taiwan, South-east Asia, Indo, PNG	Woody dicots, palms
■'Clade A.3'	Neotropics, Florida	Woody dicots, palms
■'Clade A.2'	South-east Asia	Woody dicots, palms
• *G. australe* complex B	South Africa, Australia, New Zealand, South America	

Continued

Table 1.1. Continued.

Group	Geographic categories	Hosts
• *G. austral* complex C	India, Pakistan, Taiwan, South-east Asia, New Zealand, South America	Woody dicots
Unclassified		
G. applanatum B	Europe, North America	Woody dicots
Ganoderma sp.	South Africa, South-east Asia	Woody dicots
Ganoderma sp.	Neotropics	Woody dicots
Ganoderma tsunodae (*Trachyderma*)	Japan	Woody dicots
Ganoderma colossum	South Africa, South-east Asia, Neotropics, Florida	Woody dicots

Notes: Names in parentheses are commonly miss-applied names (in 'quotes'), synonyms (=) or possible alternative names. Geographic categories and samplings are as follows: 'South Africa' includes collections from South Africa and Zimbabwe; 'Europe' includes collections from the UK, Norway, France, the Netherlands, Belgium, Austria and Germany; 'China' includes collections from mainland China with the exclusion of subtropical and tropical collections from Yunnan; 'South-east Asia' includes subtropical and tropical collections from Yunnan, Thailand, Vietnam, the Philippines, Peninsular Malaysia, Sabah and Singapore; 'Indo, PNG' includes collections from Bali, Maluku and Papua New Guinea; 'South America' includes collections from Argentina and Chile; 'Neotropics' includes collections from Costa Rica, Puerto Rico, Ecuador and French Guyana.

1.5 *Ganoderma* Pathogenicity

Ganoderma spp. produces enzymes that degrade the cell wall components of plants, lignin, cellulose and hemicellulose. The enzymes consist of polyphenol oxidases, laccases and tyrosinases that are able to degrade lignin. The fungus also produces some cellulose-degrading enzymes such as β-glucosidases (exo and endo) and hemicellulose-degrading enzymes (α-D-galactosidase, endo-β-D-mannanase, exo-β-D-mannanase and exo-β-D-mannan cellobiohydrolase). In addition, *Ganoderma* also produces amylases, extra-cellular oxidases, invertases, coagulase, protease, pectinases and cellulases (Susanto, 2002).

Ganoderma becomes infectious when the inoculum size reaches a certain level. *Ganoderma* pathogenicity can be demonstrated by artificial exposure of oil palm seedling roots to a massive inoculum source (Idris *et al.*, 2006; Breton *et al.*, 2010).

1.6 Management Control of Basal Stem Rot

Some protocols have been developed and practised to reduce the spread of *Ganoderma* disease in oil palm plantations. The control techniques are grouped into chemical, biological, agronomic and breeding controls.

Chemical control

Fungicides are used for living oil palm stands. The effectiveness of this approach is little studied but may be developed. Many fungicides have been shown to be effective in supressing the development of *Ganoderma* in laboratory conditions. However, in the field, the effectiveness of fungicides is poor (Susanto, 2002; Hushiarian *et al.*, 2013). Methods of field application may require improvement.

Biological control

Many *Trichoderma* spp. are soil-borne fungi with known antagonism to many pathogens, one of which is *Ganoderma*. The effectiveness of *Trichoderma* has been demonstrated not only in the laboratory but also in nursery and field conditions. *Trichoderma* application to the soil has been implemented by some plantation companies. In Indonesia, the application of *Trichoderma virens* to planting holes has been conducted routinely in oil palm replanting schemes since 2000 (Priwiratama and Susanto, 2014). Soil applications of *Trichoderma* are expected to reduce the incidence of *Ganoderma*.

Agronomic practice

There are some agronomic techniques that suppress *Ganoderma* disease, such as: soil mounding; surgery; sanitation; removal of diseased material; ploughing; harrowing; fallowing; planting legume cover crops; and fertilizer application. However, with the exception of fallow periods, the effectiveness of these practices is questionable and under review (Hushiarian *et al.*, 2013). Fallow periods do appear to produce a significant reduction in later *Ganoderma* incidence of the disease, but the length of time of a fallow (1 year or more) and its effectiveness are still under investigation.

Breeding

The three previous approaches aim to reduce the incidence of disease. However, it is believed that the disease incidence can be reduced significantly and infection appearance retarded if new resistant oil palm varieties are bred. Breeding programmes for *Ganoderma* resistance have been set up in Indonesia and Malaysia. Success would be a major advance in improving the integrated pest management to BSR disease (Susanto, 2002; Breton *et al.*, 2010; Purba *et al.*, 2012), but as yet little is known about the genes involved and potent resistant sources for breeding. Some seed producers in Indonesia and Malaysia claim to have planting material showing some degree of resistance (PT Socfin Indonesia (SOCFINDO), Indonesia Oil Palm Research Institute (IOPRI) and Felda Agricultural Services Sdn Bhd (FASSB)) (Amri, 2014; Anon., 2016; PPKS, 2016).

1.7 Genetic Potency of *Ganoderma* Partial Resistance

The first report of potential genetic partial resistance to *Ganoderma* in oil palm was by Akbar *et al.* (1971). From a field census of two estate plantations in Indonesia, it was reported that Deli material was more susceptible to BSR disease than the material of more African origin. A more recent census by Purba *et al.* (2012) also demonstrated that African Dura germplasm was more resistant than Deli Dura. Based on these results, it was suggested that African material had retained a wider gene pool than Deli material. This is not surprising as north-west Africa is the centre of diversity of the species, and Deli Duras are highly selected. The related oil palm species, *Elaeis oleifera*, from South America, also shows good resistance to *Ganoderma*. However, there are no reports of any oil palm material having zero incidence of BSR disease.

Breton *et al.* (2010) reported that some resistant progenies from La Me populations (originating from the Ivory Coast, Africa) transmitted resistance to their offspring. Likewise, susceptible parents tend to produce more susceptible progenies. It is suggested that resistance is a multi-genic character, and so suitable breeding strategies for *Ganoderma* partial resistance will need to be considered in developing resistant populations. Mutation breeding has recently been initiated with *Ganoderma* resistance as one target (Nur *et al.*, 2018).

1.8 Screening for *Ganoderma* Partial Resistance

The initial screening for *Ganoderma* resistance was done by conducting a census of diseased palms in the field (Akbar *et al.*, 1971; Purba *et al.*, 2012). However, these field surveys were not properly based on random locations. Therefore, the census results might have been due to a bias from the locations and not the materials. Field screening with sound statistical designs were conducted in Indonesia by Sumatra Bioscience in 2000 and by the IOPRI in 2008. Although it takes several years to collect all the results from field screening, the screening can determine the correlation between *Ganoderma* resistance, yield and yield components. Purba *et al.* (2012) reported that both *Ganoderma* resistance and yield were independent. Therefore, resistant material could be either high- or low-yielding.

1.9 Early Screening for *Ganoderma* Partial Resistance/ Susceptibility in the Nursery

It has been noted in the previous section that field screening for *Ganoderma* resistance takes several years. Therefore, a quicker method is needed, especially for breeding purposes. Breton *et al.* (2006) and Idris *et al.* (2006) first reported the possibility of screening seedlings in the nursery. Massive inoculum sources are needed to infect the roots of oil palm seedlings (Fig. 1.4).

Oil palm seedling

Root

inoculum source (rubber wood block)

Fig. 1.4. Root artificial inoculation of *Ganoderma* to infect oil palm seedling.

Both papers demonstrated the use of rubber wood blocks, inoculated with a *Ganoderma* isolate, as a source of infection for seedlings grown in polybags. CIRAD (Centre de Coopération Internationale en Recherche Agronomique pour le Développement), London Sumatra, IOPRI and MPOB (Malaysian Palm Oil Board) were among the first companies developing this method (Rajanaidu *et al.*, 2012). This approach not only reduces the time for selection but also saves on land resources, as well as allowing the selected lines to be advanced for trialling and breeding. These nursery trials take 5–7 months to conclude after germinated seeds are inoculated.

References

Akbar, U., Kusnadi, M. and Ollagnier, M. (1971) Influence of the type of planting materials and of mineral nutrients on oil palm stem rot due to *Ganoderma*. *Oleagineux* 26, 527–534.

Amri, Q. (2014) PT Socfin Indonesia: Socfindo released *Ganoderma* resistant oil palm seed (PT Socfin Indonesia: Socfindo melepas benih sawit tahan *Ganoderma*). *Sawit Indonesia*. Available at: https://sawitindonesia.com/rubrikasi-majalah/hama-penyakit/pt-socfin-indonesia-socfindo-melepas-benih-sawit-tahan-ganoderma/ (accessed 20 March 2018).

Anon. (2016) Media release: FGV launch Malaysia's first *Ganoderma* tolerant oil palm planting material. Felda Press Centre. Available at: http://www.feldaglobal.com/media-release-fgv-launch-malaysias-first-ganoderma-tolerant-oil-palm-planting-material/ (accessed 20 March 2018).

Breton, F., Hasan, Y., Hariadi, Lubis, Z. and De Franqueville, H. (2006) Characterization of parameters for the development of an early screening test for basal stem rot tolerance in oil palm progenies. *Journal of Oil Palm Research* Special Issue April, 24–36.

Breton, F., Rahmaningsih, M., Lubis, Z., Syahputra, I., Setiawati, U. *et al.* (2010) Evaluation of resistance/susceptibility level of oil palm progenies to basal stem rot disease by the use of an early screening test, relation to field observations. In: Purba, A.R., Susanto, A., Suprianto, E., Samosir, Y. and Lubis, A.F. (eds) *Second International Seminar Oil Palm Diseases – Advances in* Ganoderma *Research and Management*, 31 May 2010, Yogyakarta, Indonesia. Indonesia Oil Palm Research Institute (IOPRI), Medan, Indonesia, pp. 33–62.

Cooper, R.M., Flood, J. and Rees, R.W. (2011) *Ganoderma boninense* in oil palm plantations: current thinking of epidemiology, resistance, and pathology. *The Planter* 87(1024), 515–526.

Flood, J. (2006) A review of *Fusarium* wilt of oil palm caused by *Fusarium oxysporum* f. sp. *elaeidis*. *Phytopathology* 96(6), 660–662.

Flood, J., Hasan, Y., Turner, P.D. and O'Grady, E.B. (2000) The spread of *Ganoderma* from infective sources in the field and its implications for management of the disease in oil palm. In: Flood, J., Bridge, P.D. and Holderness, M. (eds) Ganoderma *Diseases of Perennial Crops*. CAB International, Wallingford, UK, pp. 101–112.

Hama-Ali, E.O., Panandam, J.M., Soon, G.T., Alwee, S.S.R.S., Tan, J.S. *et al.* (2014) Association between basal stem rot disease and simple sequence repeat markers in oil palm, *Elaeis guineensis* Jacq. *Euphytica* 202(2), 199–206.

Hushiarian, R., Yusof, N.A. and Dutse, S.W. (2013) Detection and control of *Ganoderma boninense*: strategies and perspectives. *Springer Plus* 2(555), 1–12.

Idris, A.S., Kushairi, D., Ariffin, D. and Basri, M.W. (2006) Technique for inoculation of oil palm germinated seeds with *Ganoderma*. *MPOB Information Series* 314, 321–324.

Jing, J.C., Seman, I.A. and Zakaria, L. (2015) Mating compatibility and restriction analysis of *Ganoderma* isolates from oil palm and other palm hosts. *Tropical Life Sciences Research* 26(2), 45–47.

Kushairi, D. (2012) Welcome remarks. In: Kien, W.C. (ed.) *Proceeding of 2012 The International Society for Oil Palm Breeders (ISOPB) International Seminar on Breeding for Oil Palm Disease Resistance*, 21–24 November 2012, Bogota, Colombia.

Lyod, A.L., Smith, J.A., Richter, B.S., Blanchette, R.A. and Smith, M.E. (2017) The laccate *Ganoderma* of the Southeastern United States: a cosmopolitan and important genus of wood decay fungi. EDIS, IFAS Extension, University of Florida, pp. 1–6.

Merciere, M., Boulord, R., Carasco-Lacombe, C., Klopp, C., Lee, Y.P. *et al.* (2017) About *Ganoderma boninese* in oil palm plantations of Sumatra and peninsular Malaysia: ancient population expansion, extensive gene flow and large scale dispersion ability. *Fungal Biology* 121(6–7), 529–540.

Moncalvo, J.M. (2000) Systematics of *Ganoderma*. In: Flood, J., Bridge, P.D. and Holderness, M. (eds) Ganoderma *Diseases of Perennial Crops*. CAB International, Wallingford, UK, pp. 23–45.

Nur, F., Forster, B.P., Osei, S.A., Amiteye, A., Ciomas, J. *et al.* (2018) *Mutation Breeding in Oil Palm: A Manual. Techniques in Plantation Science*. Forster, B.P. and Caligari, P.D.S. (eds). CAB International, Wallingford, UK (in press).

Nuranis, I., Kamaruzaman, S., Khairulmazmi, A., Mohd Shukri, I., Zulkifli, H. and Idris, A.S. (2016) Leaf nutrient status in relation to severity of *Ganoderma* infection in oil palm seedlings artificially infected with *Ganoderma boninense* using root inoculation technique. *Oil Palm Bulletin* 72, 25–31.

PPKS (2016) The construction of PPKS *Ganoderma* moderate resistant plant material (Konstruksi bahan tanaman moderat tahan *Ganoderma* PPKS). *PPKS Note*, December 2016.

Priwiratama, H. and Susanto, H. (2014) Utilization of fungi for the biological control of insect pests and *Ganoderma* disease in the Indonesian oil palm industry. *Journal of Agricultural Science and Technology* A 4, 103–111.

Purba, A.R., Setiawati, U., Rahmaningsih, M., Yenni, Y., Rahmadi, H.Y. and Nelson, S. (2012) Indonesia's experience of developing *Ganoderma* tolerant/resistant oil palm planting material. In: Kien, W.C. (ed.) *Proceeding of 2012 The International Society for Oil Palm Breeders (ISOPB) International Seminar on Breeding for Oil Palm Disease Resistance*, 21–24 November 2012, Bogota, Colombia, Paper 6, pp. 1–22.

Rajanaidu, N., Kushairi, A., Din, A., Noh, A., Norziha, A. and Ainul, M. (2012) Breeding materials for disease resistance in oil palm – current status. In: Kien, W.C. (ed.) *Proceedings of 2012 International Society for Oil Palm Breeders (ISOPB) International Seminar on Breeding for Oil Palm Disease Resistance*, 21–24 November 2012, Bogota, Colombia, Paper 1, pp. 1–2.

Rakib, M.R.M., Bong, C.F.J., Khairulmazmi, A. and Idris, A.S. (2014) Genetic and morphological diversity of *Ganoderma* species isolated from infected oil palms (*Elaeis guineensis*). *International Journal of Agriculture and Biology* 16(4), 691–699.

Rakib, M.R.M., Bong, C.F.J., Khairulmazmi, A. and Idris, A.S. (2015) Aggressiveness of *Ganoderma boninense* and *G. zonatum* isolated from upper and basal stem rot of oil palm (*Elaeis guineensis*) in Malaysia. *Journal of Oil Palm Research* 27(3), 229–240.

Rees, R.W., Flood, J., Hasan, Y., Wills, M.A. and Cooper, R.M. (2012) *Ganoderma boninense* basidiospores in oil palm plantations: evaluation of their possible role in stem rots of *Elaeis guineensis*. *Plant Pathology* 61, 567–578.

Seo, G.S. and Kirk, P.M. (2000) Ganodermataceae: nomenclature and classification. In: Flood, J., Bridge, P.D. and Holderness, M. (eds) Ganoderma *Diseases of Perennial Crops*. CAB International, Wallingford, UK, pp. 3–22.

Setiawati, U., Rahmaningsih, M., Ritonga, H., Nelson, S. and Caligari, P.D.S. (2014) Advances in breeding for *Ganoderma* partial resistance in field and nursery trials. In: Hasibuan, H.A., Rahmadi, H.Y., Amalia, R., Priwiratama, H., Wening, S., Sujadi, Sutarta, E.S., Siahaan, D. and Herawan, T. (eds) *Agriculture Proceeding of 2014 International Oil Palm Conference: Green Palm Oil for Food Security and Renewable Energy*, 17–19 June 2014, Bali, Indonesia. IOPRI, Medan, Indonesia, pp. 400–411.

Singh, G. (1991) *Ganoderma* – the scourge of oil palms in the coastal areas. *The Planter* 67, 421–444.

Steyaert, R.L. (1967) Les *Ganoderma* palmicoles. *Bulletin du Jardin Botanique National de Belgique* 37, 465–492.

Susanto, A. (2002) Biological control of *Ganoderma boninense* Pat., the causal Agent of Basal Stem Rot Disease of Oil Palm. Doctoral Dissertation. Bogor Agricultural University, Bogor.

Torres, G.A., Sarria, G.A., Martinez, G., Varon, F., Drenth, A. and Guest, D.I. (2016) Bud rot caused by *Phytophthora palmivora*: a destructive emerging disease of oil palm. *Phytopathology* 106, 320–329.

Turner, P.D. (1981) *Oil Palm Diseases and Disorders*. Oxford University Press, Oxford, 280 pp.

Wong, L.C., Bong, C.F.J. and Idris, A.S. (2012) *Ganoderma* species associated with basal stem rot disease of oil palm. *American Journal of Applied Science* 9(6), 879–885.

Zaremski, A., Lecoeur, E., Breton, F. and De Franqueville, H. (2013) Molecular characterization of fungal biodiversity and early identification of fungi associated with oil palm decay, particularly *Ganoderma boninense*. *Pro Ligno* 9(2), 3–21.

Health and Safety Considerations   **2**

Abstract

Standard and safety protocols are needed in all screening activities, both in the laboratory and in the nursery. The official standards can vary depending on country and local regulations, but good standards should be maintained whatever requirements are needed officially. The identification and elimination of hazards and risks, followed by developing specific safety procedures and procedures for preventing and responding to workplace accidents and injuries, are important features in establishing an effective occupational health and safety programme. Guidelines in health and safety issues relating to nursery screening of *Ganoderma* disease in oil palm are given below.

2.1   Health and Safety in the Laboratory

Various laboratory procedures are conducted in producing *Ganoderma* inoculum. These consist of isolating and culturing fungal isolates and preparing the inoculum. Therefore, good laboratory practices are required.

- Laboratory coats must be worn when entering the laboratory and removed when leaving the laboratory. This will give protection for the workers and the samples from outside contamination, and also people outside from laboratory contamination.
- Hands should be washed before entry and on leaving the laboratory.
- Face masks are required when working with ethanol and contaminated samples.
- Gloves are required when working with hazardous chemicals and hot, heavy and hard materials.
- Wear appropriate clothing; skin exposure should be minimized. Field clothing should be removed before entering the laboratory.

- Training should be undertaken in inoculating and flaming techniques.
- Training should be undertaken in operating an autoclave.
- Be aware of emergency procedures: firefighting, emergency exits, emergency telephone numbers, and the location of fire extinguishers, emergency shower, eyewash and first aid/first aiders.
- Be aware of hazards relating to the chemicals used in the laboratory and their Material Safety Data Sheets (MSDS information is available on the Internet), which provides information on health and safety, first aid, fire, explosion risks, disposal, how to clean up spillage, handling and storage (ScienceLab, 2013a,b,c).
- Flow cabinets (horizontal flow) should always be used when handling cultures along with their appropriate protocols.
- Be aware of standard operating procedures (SOPs) that have been developed for your laboratory, or which should be developed, for example for waste disposal, such as plastic bags, etc.

2.2   Health and Safety in the Nursery

Although the work risk in the nursery is not as high as in the field, the health and safety of the workers is paramount, and attention should always be paid to environmental considerations.

Equipment needed in the *Ganoderma* seedling screening nursery:

- Hoe for soil preparation.
- Sharp knife for cutting the seedlings for internal observation.
- Saw for cutting bamboo (bamboo is often used as poles to support the nursery roof netting).

Safety equipment that needs to be worn in the nursery:

- Rubber boots.
- Gloves for handling sharp or hard materials.
- Masks if working with hazardous chemicals such as pesticides.

Training and refreshment training are essential for worker safety. Other considerations are:

- Chemical use (especially pesticides).
- Standard operating procedures (SOPs).
- Working alone.
- Emergency procedures, first aid box.
- Be aware of nuisance insects (e.g. mosquitoes) and other animals (e.g. snakes).

Further helpful advice and guidance on health and safety in the laboratory, greenhouse and nursery can be found in Barker (2005) and Anon. (2012).

References

Anon. (2012) *Health and Safety for Greenhouses and Nurseries.* Workers' Compensation Board of British Columbia, Canada, 120 pp.

Barker, K. (2005) *At the Bench: A Laboratory Navigator.* Cold Spring Harbor Press, New York.

ScienceLab (2013a) Chloramphenicol MSDS. Material Safety Data Sheet. Available at: http://www.sciencelab.com/msds.php?msdsId=9927131 (accessed 21 March 2018).

ScienceLab (2013b) Ethyl Alcohol 190 Proof MSDS. Material Safety Data Sheet. Available at: http://www.sciencelab.com/msds.php?msdsId=9923956 (accessed 21 March 2018).

ScienceLab (2013c) Streptomycin sulfate MSDS. Material Safety Data Sheet. Available at: http://www.sciencelab.com/msds.php?msdsId=9927414 (accessed 21 March 2018).

Media Preparation for *In Vitro* Culture of *Ganoderma*

3

Abstract

The preparation of growth media is the first step in isolating any micro-organism (bacteria and fungi) from the environment. Media may be solid or liquid. The culture medium formulation will affect the success of mycelium growth in culture. Although there are many culture media described for fungi, only a few of them have proven to be excellent in culturing *Ganoderma*. This chapter explains how to prepare the different media used in culturing *Ganoderma* isolates that cause BSR disease. The methods provided here are for 1 l of medium (Tables 3.1 and 3.2), which may be scaled up or down depending on capacity and the volumes required; a table is provided giving the volumes of media needed with respect to the required numbers of cultures (see Chapter 5 of this manual, Table 5.5).

3.1 Water Agar Medium

Water agar (WA) is the simplest agar medium (Himedia, 2015). WA is often recommended as it is cheap and supports spore germination collected from *Ganoderma* sporophores. Any sample from any part of the fungus can be used to isolate cultures using this medium. Two antibiotics are deployed in the medium to prevent the growth of bacterial contaminants. Typically, growth on WA medium is relatively slow. This is excellent, because success in fungus isolation can be achieved by the use of selective media that slow down the growth of the fungus.

Methods for 1 l of WA

1. Before starting the preparation, all equipment and workplaces must be checked. Prepare all materials needed based on requirements.

© Miranti Rahmaningsih, Ike Virdiana, Syamsul Bahri, Yassier Anwar, Brian P. Forster and Frédéric Breton 2018. *Nursery Screening for Ganoderma Response in Oil Palm Seedlings: A Manual*

Table 3.1. Materials needed for 1 l of medium.

Materials	Water agar (WA)	*Ganoderma* selective medium (GSM)	Potato dextrose agar (PDA)
		Medium (1 l)	
Agar-agar bars/agar-agar powder	20 g	11 g (Mix A)	15 g
Bacto peptone		5 g (Mix A)	
Benlate T-20		0.15 g (Mix B)	
Chloramphenicol	0.5 g	0.1 g (Mix B)	0.5 g
Distilled water	1000 ml	900 ml (Mix A); 80 ml (Mix B)	1500 ml
Ethanol 70%	As necessary	As necessary	As necessary
Ethanol 95%		20 ml (Mix B)	
Lactic acid 50%		2 ml (Mix B)	
Magnesium phosphate		0.25 g (Mix A)	
Pentachloronitrobenzene (PCNB)		0.285 g (Mix B)	
Potassium phosphate (K_2HPO_4)		0.5 g (Mix A)	
Potato			200 g
Ridomil 25 WP		0.13 g (Mix B)	
Streptomycin sulfate	0.5 g	0.3 g (Mix B)	
Sucrose			20 g
Tannic acid powder		1.25 g (Mix B)	

Table 3.2. Check list of equipment and tools needed for media preparation.

Equipment and tools	Water agar (WA)	*Ganoderma* selective medium (GSM)	Potato dextrose agar (PDA)
		Medium	
Aluminium foil	√	√	√
Analytical balance	√	√	√
Autoclave	√	√	√
Boiling pan			√
Bunsen burner	√	√	√
Erlenmeyer flask	√	√	√
Laminar flow cabinet (horizontal)	√	√	√
Measuring glass	√	√	√
Plastic container	√	√	√
Sieve (mm gauge)			√
Sterile Petri dish	√	√	√
Stirring hot plate	√	√	√
Stove			√

2. Mix all the materials (Table 3.1, except for antibiotics, chloramphenicol and streptomycin sulfate) by putting them into an Erlenmeyer flask and stir with heating for 20 min (± 5 min). The solid materials should dissolve when mixed with the liquids.

3. Sterilization is conducted by autoclaving at 0.11 MPa pressure (121°C) for 15 min. Both horizontal and vertical autoclaves can be used to sterilize the media. Since autoclaves use high pressure and high temperature, work instructions for operating autoclaves must be strictly adhered to and training should be given to all potential operators (see Chapter 2 of this manual).

4. Clean all laminar work surfaces with a disinfectant solution prior to use and leave switched on for at least 5 min before use. It is also essential to reduce the airflow as much as possible during the transfer of plates to avoid contamination.

5. The media are cooled in a laminar flow cabinet until the temperature reaches around 40°C. Then, an antibiotic solution, previously dissolved in 5 ml of distilled water, is added to the media using a sterile syringe connected to a 0.2 μm filter. The media are stirred slightly before use.

6. The hot medium is poured into sterile Petri dishes. A Bunsen burner is used to flame the mouth of the Erlenmeyer flask before the medium is poured out into Petri dishes (Fig. 3.1). This activity is conducted in a laminar flow cabinet. The medium will set after 30–60 min at room temperature. Store the medium in clean plastic containers in a room which is below 30°C.

3.2 *Ganoderma* Selective Medium

Another medium used for the first step in *Ganoderma* isolation is the *Ganoderma* selective medium (GSM). The first publication of this medium was by Ariffin and Idris in 1991. In addition to using antibiotics, this medium also contains a fungicide to prevent the growth of fungal contaminants. The effectiveness of GSM is similar to WA agar in isolating *Ganoderma*.

Methods for 1 l of GSM

1. Before starting the preparation, all equipment and workplaces must be checked. Prepare all materials needed based on requirements.

2. Put all Mix A (Table 3.1) in an Erlenmeyer flask and then mix by stirring on a hotplate until all the solid materials are mixed and dissolved with the liquid materials. Sterilize the medium under 0.11 MPa pressure (121°C) for 15 min. The next activities are conducted in a laminar flow cabinet.

3. Clean all work surfaces with a disinfectant solution. It is essential to reduce the airflow as much as possible during the transfer of plates to avoid contamination. A Bunsen burner is used to flame the mouth of the Erlenmeyer

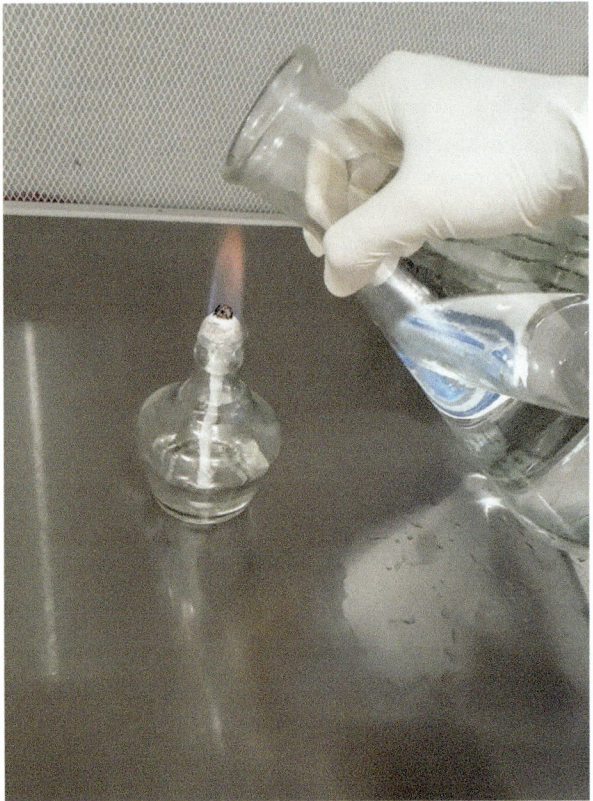

Fig. 3.1. Flaming the mouth of Erlenmeyer flask to sterilize before pouring the medium; training is required to learn this skill.

flask before the medium is poured into (Fig. 3.1) sterile Petri dishes. This is a skilled operation for which training is needed. Gloves are generally not worn as the operator needs to feel the temperature of the flask, which should not become too hot.

4. In the laminar flow cabinet, put all the materials of Mix B (Table 3.1) in an Erlenmeyer flask and mix by stirring on a hotplate, followed by mixing both solutions (Mix A and Mix B) in one Erlenmeyer flask. When all materials are mixed, pour them into sterile Petri dishes. The medium will set after 30–60 min at room temperature. The medium can be stored in clean plastic containers in a media storage room below 30°C.

3.3 Potato Dextrose Agar Medium

WA and GSM media are commonly used in the first *Ganoderma* isolation steps, while potato dextrose agar (PDA) is used for the culture of the fungus after isolation (Himedia, 2016). The growth of the fungus is more rapid and dense on PDA. This is because PDA contains more carbohydrates (from potato and sucrose). This medium generally also contains less antibiotics, to allow the fungus to grow rapidly and densely.

Methods for 1 l of PDA (Table 3.1) medium

1. Before starting the preparation, all equipment and workplaces must be checked. Prepare all the materials needed based on requirements. Potato tubers should be peeled before being weighed.
2. To prepare potato extract, boil 200 g sliced potato in 1.5 l of distilled water for about 20 min (± 5 min) (Fig. 3.2). The boiled potato pieces are cooled and sieved before being mixed with the other ingredients.
3. The potato extract is mixed with other materials (except for antibiotics, chloramphenicol and streptomycin sulfate) in an Erlenmeyer flask and stirred on a hotplate for 20 min (± 5 min). The solid materials should be dissolved and mixed with the liquid preparation prior to autoclaving. An Erlenmeyer flask is prepared, together with its stopper (consisting of absorbent cotton and absorbent gauze) and aluminium foil.
4. Sterilization of the medium is conducted by autoclaving at 0.11 MPa pressure (121°C) for 15 min. Both horizontal and vertical autoclaves can be used to sterilize the medium. Since the autoclave uses high pressure and high temperature, the instructions for operating the autoclave must be strictly adhered to and training should be given to the workers (see Chapter 2 of this manual).
5. The preparation of the laminar flow bench, including blower and contamination checks, should be conducted prior to any activities under laminar flow. Clean all work surfaces with a disinfectant solution. It is also essential to reduce external airflow as much as possible during the transfer of plates, to avoid contamination.
6. Media are cooled in the laminar flow cabinet until the temperature reaches around 40°C. Then, antibiotic solution previously dissolved in 5 ml

Fig. 3.2. Preparation of potato extract: (a) peeled and cut potatoes; (b) cut potatoes are boiled.

distilled water is added to the media using sterile syringe connected to a 0.2 μm filter. The media preparations are stirred slightly before use.

7. The hot medium is poured into sterile Petri dishes. A Bunsen burner is used to flame the mouth of the Erlenmeyer flasks before the medium is poured into Petri dishes (Fig. 3.1). The medium will set after 30–60 min at room temperature after pouring. The medium can be stored in a clean room below 30°C.

References

Ariffin, D. and Idris, A.S. (1991) A selective medium for the isolation of *Ganoderma* from diseased tissues. In: Yusof, B. (ed.) *Proceedings of the 1991 International Palm Oil Conference; Progress, Prospects*, and *Challenges towards the 21st Century (Model I, Agriculture)*, 9–14 September 1991. Palm Oil Research Institute of Malaysia, Bangi, Selangor, Malaysia, pp. 517–519.

Himedia (2015) *Water Agar*. Technical Data. Himedia Laboratories, Mumbai, India.

Himedia (2016) *Potato Dextrose Agar*. Technical Data. Himedia Laboratories, Mumbai, India.

Collecting *Ganoderma* Isolates and Culture Preparation

4

Abstract

Ganoderma may be isolated from the basidiocarp (the sporophore of basidiomycete), infected palm tissues (trunk or roots) or from single basidiospores; these sources produce abundant *Ganoderma* mycelia in culture (Ho and Nawawi, 1986). Water agar (WA) or *Ganoderma* selective medium (GSM) is used as the first growth medium; later, potato dextrose agar (PDA) is used for rapid multiplication of *Ganoderma* cultures. The selection of samples especially from the basidiocarp can affect the success of *Ganoderma* isolation; the use of younger basidiocarps is more favourable than older ones. An aseptic environment is needed for the successful isolation and culturing of *Ganoderma*. Checklists of materials and equipment needed are provided in Tables 4.1 and 4.2.

4.1 Dikaryotic Mycelium Isolation from Basidiocarps

The use of young basidiocarps is preferred as they have more active mycelium than older basidiocarps. Also, the old basidiocarps have very thin dikaryotic mycelium, which is harder to isolate (Fig. 4.1). The difference between young and old basidiocarps can be seen by their colour; old basidiocarps usually have a darker colour than young basidiocarps. Moreover, the 'hat' of old basidiocarps is harder and can show insects within, leading to more difficulty in isolation work.

Methods

1. *Ganoderma* basidiocarps are harvested from infected palms (basal stem rot (BSR) or upper stem rot (USR) palms). If the basidiocarps are from USR incidence (basidiocarps are at the top of the palm), a chainsaw is needed to

© Miranti Rahmaningsih, Ike Virdiana, Syamsul Bahri, Yassier Anwar, Brian P. Forster and Frédéric Breton 2018. *Nursery Screening for Ganoderma Response in Oil Palm Seedlings: A Manual*

Table 4.1. Checklist of materials needed for *Ganoderma* isolation.

Materials	Isolation samples		
	Basidiocarps	Infected tissues	Basidiospores
Distilled water	√	√	√
70% ethanol	√	√	√
Ganoderma basidiocarps	√		
Ganoderma basidiospores			√
Infected tissues		√	
PDA	√	√	√
WA or GSM	√	√	√
Water media	√	√	√

Notes: GSM = *Ganoderma* selective medium; PDA = potato dextrose agar; WA = water agar.

Table 4.2. Checklist of equipment and tools needed for *Ganoderma* isolation.

Materials	Inoculation samples		
	Basidiocarps	Infected tissues	Basidiospores
Aluminium foil	√	√	√
Beaker-glass	√	√	√
Bunsen burner	√	√	√
Cell tape or Parafilm	√	√	√
Chainsaw	√ (USR)	√	
Forceps	√	√	
Haemocytometer			√
Inoculation blade	√	√	√
Inoculation loop			√
Knife	√	√	
Laboratory filter paper			√
Laminar flow cabinet (horizontal)	√	√	√
Marker pen	√	√	√
Microscope			√
Plastic bag	√	√	√
Scissors			√
Sterile Petri dish	√	√	√

Note: USR = upper stem rot.

fell the palm. The chainsaw should be used by a trained worker and safety precautions taken (see Chapter 2 of this manual). It is better to harvest more than one basidiocarp from an infected palm, with an aim to check whether or not the basidiocarps are from the same individual. This checking can be made through *Ganoderma* seedling screening by looking at their infection characteristics or by DNA profiling (genotyping).

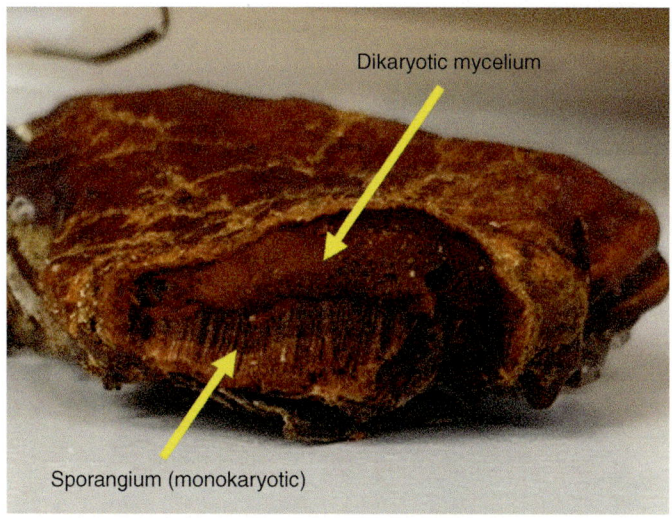

Dikaryotic mycelium

Sporangium (monokaryotic)

Fig. 4.1. *Ganoderma* basidiocarp consists of dikaryotic mycelium and sporangium (monokaryotic).

2. In the laboratory, the basidiocarps are cut into small cubes (about $1 \times 1 \times 1$ cm). The dikaryotic mycelium part of the basidiocarp is targeted for isolation purposes. It is located above the sporangium (see Fig. 4.1). Surface sterilization is conducted by washing the mycelium pieces with distilled water and then dipping in 70% ethanol for approximately 1 s.

3. The sample is cut again into smaller pieces (about $5 \times 5 \times 5$ mm), and each piece placed into WA media or GSM (4–5 pieces/plate). Label: write the isolate's name, isolation date, sample code and medium's name on the surface of the plate.

4. Culture conditions must be free from contaminants (other fungi and/or bacteria). The inoculation operator must maintain personal cleanliness to avoid contamination.

5. Wait until there is mycelium growing around the cultured samples. The time taken for mycelium to appear depends on the samples. Old samples will take longer than new samples. Normally, it takes 2 weeks to 1 month from the initial culture of fresh basidiocarp samples. From this mycelium culture, take out a slice on medium that carries the longest mycelium growth from each sample. It is expected that the most distant part will have the least contamination. Place the slice in WA or on to PDA medium. For the next re-culture, it is suggested that PDA medium is used. All the fungi are incubated at room temperature $28 \pm 1°C$ and in the dark.

6. For long storage or conservation, the isolates (solid medium containing the mycelium of fungus) are placed in a Wheaton tube (8 ml) with 5 ml of sterile distilled water and stored in an incubator at 20°C. Cultures may be kept for more than 1 year.

4.2　Isolation from Infected Tissues

The isolation from infected tissues is slightly different from that of basidio-carps. A sample is taken from a palm part where there are both infected tissues and healthy tissues (Fig. 4.2). It is expected that the mycelium from the infected part is active in healthy neighbouring tissues. Therefore, the growing mycelium can be isolated on agar media.

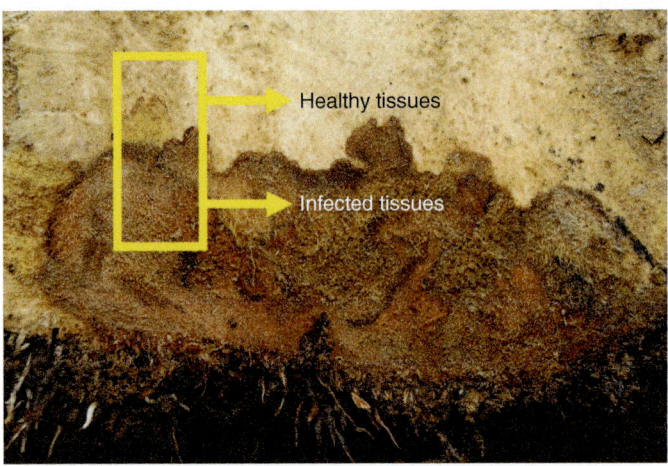

Fig. 4.2. Infected trunk consists of both infected tissues and healthy tissues.

Methods

1. The infected tissues can be taken from both fallen or standing palms as long as the palm is infected by *Ganoderma*. If the palm has already fallen, cut the trunk or roots using a chainsaw until healthy and infected neighbouring tissues are located. The chainsaw should be used by a trained worker and appropriate safety precautions taken (see Chapter 2 of this manual).

2. In the laboratory, the samples with infected parts and healthy parts are cut into small pieces (about $1 \times 1 \times 1$ cm). Surface sterilization is conducted by soaking the tissue samples in water, followed by dipping them in 70% ethanol for approximately 1 s for each.

3. Re-cut the tissue samples into smaller pieces (about $5 \times 5 \times 5$ mm), then place the samples on to WA medium or GSM (4–5 pieces/plate) (Fig. 4.3). Label: write the isolate's name, isolation date, sample code and medium's name on the surface of the plate. The culture must be free from any contaminants (other fungi and/or bacteria).

4. Wait until there is mycelial growth around the samples. From this mycelium, take one slice of medium containing mycelia with the longest growth; it

Fig. 4.3. *Ganoderma* isolation from infected oil palm tissues.

is expected that the most distant part has less contaminants than short mycelia. Place the slice on to WA or PDA medium. For the next re-culture, it is suggested that PDA medium is used to produce vigorous cultures. All the fungi are incubated at room temperature $28 \pm 1°C$ and in the dark.

5. For long storage (more than 1 year) or conservation, the isolates may be placed in a Wheaton tube (8 ml) with 5 ml of sterile distilled water. Store the isolates in an incubator at 20°C. Cultures may be kept for more than 1 year.

4.3 Single Basidiospore Isolation

The isolation of *Ganoderma* basidiospores is more difficult compared to the two methods above (Sections 4.1 and 4.2). The germination of a single spore often fails. Inappropriate culture medium selection and the concentration of spores are two factors that often cause failure in isolation. The timing of spore harvesting also affects the number of spores and their viability. The best time to harvest *Ganoderma* basidiospores is in the evening.

Methods

1. Basidiospores are collected from the basidiocarp of *Ganoderma* by trapping them in Whatman filter paper covered by aluminium foil (Fig. 4.4). This trapping method allows prolonged and more sensitive spore sampling in the field compared to traditional traps using Petri dishes containing a selective medium. The best time to harvest spores is in the evening. Traps may be left for 1 day.

2. Cut the filter paper into small (1×1 cm) pieces and soak with sterile water. The concentration of spores is determined by observation at $100\times$ magnification under a light microscope using a haemocytometer. The spore concentration should be about 2000 spores/ml of water. This concentration can be performed by a serial dilution technique.

3. $20\ \mu l$ of solution are spread on the surface of WA medium or GSM by using an inoculating loop. Label: write the isolate's name/code, isolation

Fig. 4.4. (a) Basidiocarp on an infected palm trunk; (b) basidiocarp is wrapped with filter paper covered in aluminium foil; (c) spores (ringed) are trapped and harvested on the filter paper.

date, sample code and medium's name on the surface of the plate. It is suggested that a methodical system is adopted when pouring the medium to track the germination of the spores.

4. After 24–48 h of incubation in the dark at $28 \pm 1°C$, spore germination is observed with a microscope (under a laminar flow cabinet) and some germinated spores are individually collected using a sterile needle and placed in a PDA medium.

5. For the next re-culture, it is suggested that PDA medium is used. The cultures are incubated in the dark at a room temperature of $28 \pm 1°C$. For long storage (more than 1 year) or conservation, the isolates are placed in a Wheaton tube (8 ml) with 5 ml of sterile distilled water and stored below 20°C. Cultures may be kept for more than 1 year.

4.4 *Ganoderma* Culture Preparation

WA, GSM and PDA media can be used for culturing *Ganoderma* isolates. However, to have more fungi biomass, the use of PDA is preferred. Other options include yeast malt agar (YMA) and mushroom complete medium (MCM). The materials, equipment and tools needed for *Ganoderma* culture preparation are given in Table 4.3.

Table 4.3. Materials, equipment and tools needed for *Ganoderma* culture preparation.

Materials	Equipment and tools
Ganoderma pure culture (see Fig. 4.5) PDA medium	Bunsen burner Cell tape or Parafilm Inoculation blade Laminar flow cabinet Marker pen Sterile Petri dishes

Note: PDA = potato dextrose agar.

Methods

1. Prepare both the PDA medium and pure *Ganoderma* isolate (Fig. 4.5). Before starting to culture the isolate, all the surfaces of the equipment and tools must be cleaned with a disinfectant solution or 70% ethanol. The inoculation blade is flame sterilized using a Bunsen burner.

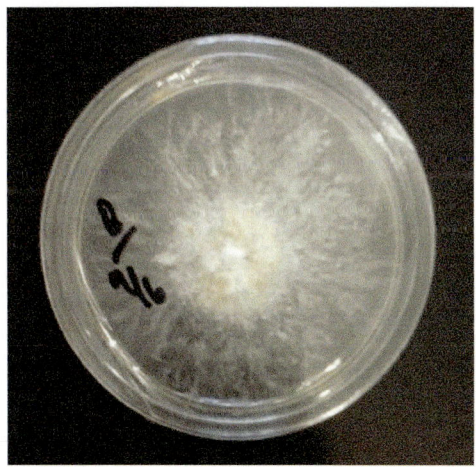

Fig. 4.5. Pure culture of *Ganoderma* mycelium.

2. Cut the pure isolate into small pieces (5 × 5 mm). Take one slice of the isolate and place it on to the PDA medium. The isolate should be placed upside down so that the mycelium is touching the medium.

3. Close and strengthen the Petri dish with tape or Parafilm. Label: write the name of the isolate, subculturing date and the name of the medium on the surface of the Petri dish. The medium can be stored in the dark at $28 \pm 1°C$. Wait for 2 weeks to 1 month until mycelium covers the surface and is ready to be used.

Reference

Ho, Y.W. and Nawawi, A. (1986) Isolation, growth, and sporophore development of *Ganoderma boninense* from oil palm in Malaysia. *Pertanika* 9(1), 69–73.

Preparation of *Ganoderma* Inoculum

5

Abstract

To conduct a nursery screen, a massive amount of *Ganoderma* inoculum is needed. This is used to inoculate/infect oil palm seedlings artificially. The success of nursery screening relies on a set of standardized parameters; one of them is the effectiveness of the inoculum source. Various parameters, such as the selection of aggressive isolates, material for the substrates, the volume of substrates and the length of fungus incubation, need to be standardized in order to provide robust and reproducible results. Currently, the use of rubber wood blocks as a substrate for *Ganoderma* is the most effective method of exposing and infecting seedlings with *Ganoderma*.

5.1 Preparation of Potato Agar Medium

In addition to the use of rubber wood blocks (RWBs) as a substrate, agar medium is also used as a starter food for *Ganoderma* to grow. This medium is added to the RWBs. It is different from potato dextrose agar (PDA), as potato is the only source of carbohydrate in potato agar (PA) and there are no antibiotics in this medium. The materials, equipment and tools required in preparing potato agar medium are given in Table 5.1.

Table 5.1. Materials, equipment and tools needed for PA preparation.

Materials (for 1 l of PA media)	Equipment and tools
9 g agar-agar bars/5 g agar-agar powder	Analytical balance
70% ethanol as necessary	Boiling pan
1.5 l distilled water	Knife
200 g potatoes	Measuring glass cylinder
	Sieve
	Stove

Note: PA = potato agar.

© Miranti Rahmaningsih, Ike Virdiana, Syamsul Bahri, Yassier Anwar, Brian P. Forster and Frédéric Breton 2018. *Nursery Screening for Ganoderma Response in Oil Palm Seedlings: A Manual*

Methods

1. Prepare all materials needed based on requirements. Potatoes should be peeled before being weighed. Cut the potatoes into small pieces and boil them for 20 min (± 5 min). Time is counted from when the water has started to boil (see Chapter 3 of this manual, Fig. 3.2).
2. Sieve the potato extract and measure the extract based on the requirements. Boil the extract again with agar (agar-agar bars or agar-agar powder).

5.2 Preparation of Rubber Wood Blocks

The use of RWBs has been observed to be the most effective substrate for *Ganoderma*. Previous studies comparing RWBs with other substrates such as liquid medium, sawdust-based and oil palm wood block proved RWBs to be the most effective (Breton *et al.*, 2006). The materials, equipment and tools needed in the preparation of RWBs are given in Table 5.2. Moreover, the availability of rubber wood is reliable, since there are often many rubber plantations in areas where oil palm is grown and many trunks are available at the time of replanting. The unproductive rubber wood from the field is sent to the sawmill (usually a small local one) to produce wood blocks based on the requirement. The size of the RWB can vary; we suggest the use of 6 × 6 × 6 cm and 6 × 6 × 3 cm blocks (Fig. 5.1). However, there is a problem with RWB waste after use. Since, in line with sustainable policies, zero burning systems are applied in all oil palm plantations, other methods of destroying the *Ganoderma* inoculum waste should be considered, e.g. incineration.

Table 5.2. Materials, equipment and tools needed for the preparation of rubber wood blocks.

Materials	Equipment and tools
PA media (50 l for each bag of RWBs)	Aluminium foil
RWB size 6 × 6 × 6 cm or 6 × 6 × 3 cm	Autoclave
	Boiling pan, diameter 50 cm, height 70 cm
	Marker pen
	Measuring glass cylinder
	Polypropylene (PP) plastic bag, size 22 × 36 cm
	PVC pipe
	Stopper from hydrophobic cotton wool and gauze
	Stove
	Wool thread

Notes: PA = potato agar; RWBs = rubber wood blocks.

Fig. 5.1. The preparation of rubber wood blocks in a sawmill.

Methods

1. Prepare RWBs based on the requirement. Boil the RWBs for 5–7 h. Time is taken from when the water starts to boil. Top up with hot water to ensure the RWBs are continually covered with boiling water (Fig. 5.2).

Fig. 5.2. Rubber wood blocks are boiled for 5–7 h.

2. After boiling, RWBs are cooled overnight while remaining immersed in the boiled water, to swell themselves with water.

3. The next working day, pack the RWBs in polypropylene (PP) plastic bags (two RWBs per polybag). Put a PVC pipe into the mouth of the plastic bag and bind it with thread.

4. Fill each bag with 50 ml (± 2 ml) of PA medium. Measure the medium with a plastic measuring cylinder. Plug the mouth of the plastic bag with a stopper and then cover it with aluminium foil (Fig. 5.3).

Fig. 5.3. (a) Rubber wood blocks are placed in a PP plastic bag and PA medium added via a PVC pipe through the mouth of the bag. (b) The mouth of the plastic bag is sealed by a stopper. (c) The mouth of the plastic bag is then covered with aluminium foil.

5. Sterilize the RWBs by autoclaving at 0.11 MPa pressure (121°C) for 1 h. After the sterilization process, cool down the RWBs. Write the trial name, isolate name and inoculation date on the plastic bag. Leave the RWBs for 1 night and then inoculate them (as noted below) the next day.

5.3 Inoculation of Rubber Wood Blocks

This process is similar to the other culture preparations. The only difference is the medium used. Culture preparation uses PDA as the medium for *Ganoderma* to grow, while this process uses RWBs as a medium substrate. The *Ganoderma* isolates to be used should have been tested previously and proven to be aggressive. An investigation from Rakib *et al.* (2015) found that an isolate of *Ganoderma zonatum* of upper stem rot (USR) was the most aggressive, followed by the isolates from *G. zonatum* and *Ganoderma boninense* of basal stem rot (BSR). The choice of species should reflect the dominant *Ganoderma* species/strain in the region. The materials, equipment and tools needed for the inoculation of RWBs are given in Table 5.3.

Table 5.3. Materials, equipment and tools needed for inoculation of rubber wood blocks.

Materials	Equipment and tools
70% ethanol	Bunsen burner
Ganoderma pure culture	Inoculation blade
Sterilized RWBs	Laminar flow cabinet (horizontal)
	Rubber bands

Note: RWBs = rubber wood blocks.

Methods

1. RWB inoculation is conducted 1 day after RWB preparation using previously prepared *Ganoderma* pure cultures.
2. Before starting to inoculate the RWBs, all surfaces of the equipment and tools must be cleaned with a disinfectant solution or 70% ethanol. Keep the inoculation blade sterile; flame it with a Bunsen burner.
3. Cut the pure isolate into small pieces (about 5×5 mm), open the aluminium foil that covers the RWBs (inside the PP plastic bag), spray the aluminium foil and stopper with 70% ethanol and then open the stopper.
4. Take six to eight slices of pure isolate, put these into a plastic bag with a RWB and spread those isolate pieces on the surfaces of the RWBs (Fig. 5.4). Plug the bags with a stopper. Wrap the bags with aluminium foil and secure them by using rubber bands. The RWBs are ready to be stored in the culture room at $28 \pm 1°C$ in the dark.

5.4   Incubation of Rubber Wood Blocks

The RWBs need to be incubated for 12–16 weeks to build up inoculum in the $6 \times 6 \times 6$ cm or the $6 \times 6 \times 3$ cm RWBs. The time needs to be determined for other sizes. The incubation room should be kept dark, except when it is essential to work in the room. The materials, equipment and tools needed for the incubation of RWBs are given in Table 5.4. The media and cultures used for routine screening are given in Table 5.5.

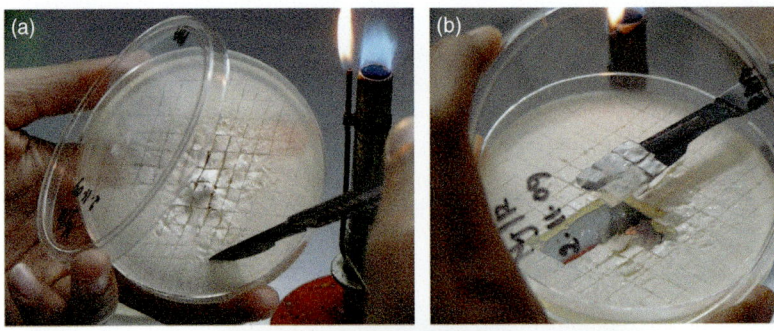

Fig. 5.4. The inoculation process of *Ganoderma* to rubber wood blocks (RWBs). (a) Pure culture of *Ganoderma* is cut into small pieces. (b) Six to eight pieces are taken to be inoculated to RWBs. (c) The pieces of *Ganoderma* pure culture are put into a plastic bag and spread on the surface of the RWBs.

Table 5.4. Materials, equipment and tools needed for incubation of rubber wood blocks.

Materials	Equipment and tools
70% ethanol	Pen
Inoculated RWBs	Trolley

Note: RWBs = rubber wood blocks.

Table 5.5. Media and cultures used for routine screening.

Materials	Quantity[a]
Rubber wood blocks	3150 blocks (3000 + (5% of 3000 for supplying))
Potato agar	± 80 l
PDA (for culture preparation)	± 8 l

Notes: [a]The quantity is based on routine screening that tests 50 progenies × (5 replicates + 1 replicate for supplying) × 10 seeds/seedlings. PDA = potato dextrose agar.

Methods

1. Place the inoculated RWBs in their bags on to a trolley and transfer them to a dark culture room/incubation room at a temperature of $28 \pm 1°C$, with air circulation (e.g. ceiling fans).

2. Put information on the screening trial name, the size of the RWBs, the RWBs' inoculation date, the name of the isolate and the field inoculation date on the rack. Wait for 12–16 weeks for the RWBs to be ready for seedling screening (Fig. 5.5).

Fig. 5.5. *Ganoderma* inoculum on rubber wood blocks sealed inside a bag.

References

Breton, F., Hasan, Y., Hariadi, Lubis, Z. and De Franqueville, H. (2006) Characterization of parameters for the development of an early screening test for basal stem rot tolerance in oil palm progenies. *Journal of Oil Palm Research* Special Issue April, 24–36.

Rakib, M.R.M., Bong, C.F.J., Khairulmazmi, A. and Idris, A.S. (2015) Aggressiveness of *Ganoderma boninense* and *G. zonatum* isolated from upper- and basal stem rot of oil palm (*Elaeis guineensis*) in Malaysia. *Journal of Oil Palm Research* 27(3), 229–240.

Nursery Inoculation

6

Abstract

Parameters critical in the nursery for successful screening include: the level of shade; the inoculation stage; and the inoculation process as well as the trial conditions. Similar to the process in the laboratory, each parameter in the nursery should be as standardized as possible. However, variation in parameters will always occur. Therefore, standardization is aimed to minimize the variation between and within trials.

6.1 Nursery Preparation

Shade level is the most important parameter during the preparation of a nursery for *Ganoderma* screening (Fig. 6.1). Previous experiments have demonstrated that 90% shade gives the highest infection of the inoculated oil palm seedlings, probably because *Ganoderma* grows rapidly under dark conditions. The ambient temperature affects the success of seedling inoculation. Although *Ganoderma* depends on soil temperature, ambient air temperature has a significant impact on the temperature inside the nursery polybag. The materials, equipment and tools needed for nursery preparation are given in Table 6.1.

Methods

1. One important step before starting a screening trial is preparation of the trial design with respect to purpose and statistical rigour.
2. Two weeks before planting, prepare plots based on the experimental design. At the same time, prepare the polybags by filling them with topsoil. The soil should be sieved to remove root and other debris. The surface of the soil should be 14 cm below the top of the polybag (Fig. 6.2).

© Miranti Rahmaningsih, Ike Virdiana, Syamsul Bahri, Yassier Anwar, Brian P. Forster and Frédéric Breton 2018. *Nursery Screening for Ganoderma Response in Oil Palm Seedlings: A Manual*

Fig. 6.1. *Ganoderma* seedling screening nursery with shading.

Table 6.1. Materials, equipment and tools needed for nursery preparation.

Materials	Equipment and tools
Soil (topsoil)	Nursery construction with net shading
	Paint
	Polybag, size $20 \times 30 \times 0.1$ cm
	Soil sieve, opening size 0.5×0.5 cm

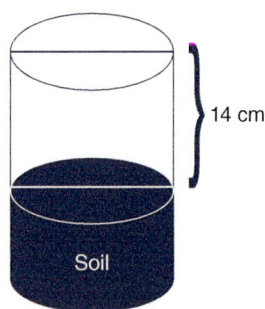

Fig. 6.2. Drawing showing the distance between the soil surface and the top edge of the polybag.

3. Place the polybags into their positions in the plot. Label the polybags with the trial number, replicate number, treatment/progeny code and the number of the polybag based on the trial map. Place a label to identify the replicate and its plot location.

6.2 Selection of Inoculated Rubber Wood Blocks

The selection/rejection of inoculated rubber wood blocks (RWBs) is done in a dark culture room and is based on the absence of contaminants (bacteria and other fungi). Inoculated RWBs with heavy or light contamination should be rejected. Training should be given to the workers in selecting RWBs for seedling screening. The materials, equipment and tools needed for selecting inoculated RWBs are given in Table 6.2.

Table 6.2. Materials, equipment and tools needed for selecting inoculated rubber wood blocks (RWBs).

Materials	Equipment and tools
Inoculated RWBs	Flashlight for inspection

Methods

1. One week before the germinated seeds are planted, select the inoculum source (RWBs inoculated with *Ganoderma* mycelium).
2. Before inoculated RWBs are placed into polybags, make sure (re-check) the RWBs are not contaminated by bacteria or other fungi.

6.3 Nursery Inoculation

Between 1 week to 1 day before planting, the inoculated RWBs are placed into the polybag. This stage is crucial because any variation will affect the success in infecting the seedlings. The distance between the surface of the RWB and the surface of the polybag (normally 14 cm) is critical and should be standardized. The materials, equipment and tools needed for nursery inoculation are given in Table 6.3.

Table 6.3. Materials, equipment and tools needed for nursery inoculation.

Materials	Equipment and tools
Polybag, $20 \times 30 \times 0.1$ cm with soil Selected inoculated RWBs	Calibration tool

Note: RWBs = rubber wood blocks.

Methods

1. Place the selected inoculated RWB into the polybag at about 8 cm from the surface of the polybag (Figs 6.3 and 6.4). A calibration tool should be used to standardize the distance between the surface of the RWB and the surface of the polybag (Fig. 6.3).

2. Cover the inoculum source with topsoil (Fig. 6.4).

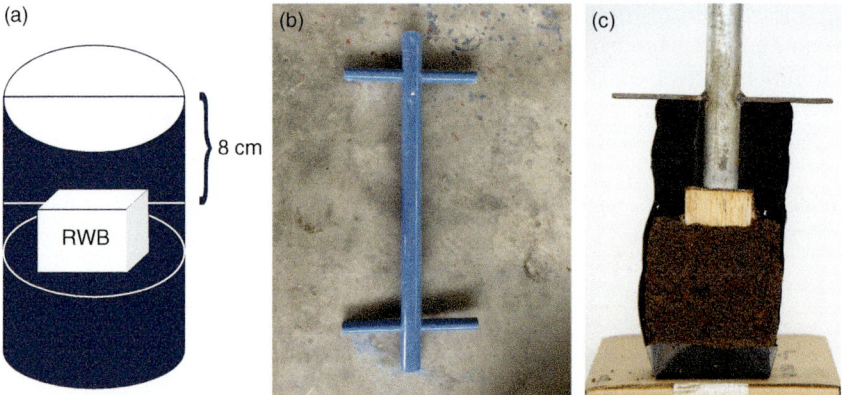

Fig. 6.3. (a) Drawing showing the distance between the surface of the RWB and the surface of the polybag; (b) calibration tool; (c) standardization of distance using a calibration tool.

Fig. 6.4. *Ganoderma* inoculation process in the nursery. (a) Polybags are filled with growing medium (soil); (b) inoculated rubber wood blocks are placed on the soil; (c) inoculated rubber wood blocks are then covered by growing medium (soil).

6.4 Planting

Planting materials that are used for nursery screening should be either 3-month-old seedlings or germinated seeds. In order to minimize root injury during this stage, the use of germinated seeds is favoured, and is simpler in that less resources are used (Fig. 6.5). The materials, equipment and tools needed for planting are given in Table 6.4.

Fig. 6.5. Longitudinal section of prepared polybag: germinated seed is planted with inoculated rubber wood block covered with soil.

Table 6.4. Materials, equipment and tools needed for planting.

Materials	Equipment and tools
Oil palm germinated seeds	Pen
Oil palm seedlings	

Methods

1. Polybags are fully watered 1 day before planting.
2. On the day of planting, receive the planting materials complete with information. Record the name of the treatment/progeny code and the

material/cross details. Check that the information on the label attached to the seed bag matches that of the received seedlings/germinated seed.
3. Plant oil palm seedlings (3 months old) or germinated seeds in polybags based on the 'trial map' or experimental design. Record the trial number, the inoculation date, the planting date, the name of the isolate, the progeny code and the size of the rubber wood block on a plate in front of the trial plots.

Scoring Response to *Ganoderma*   **7**

Abstract

To assess the variation in *Ganoderma* resistance, two different observations are conducted in the nursery. The scoring system of *Ganoderma* nursery screening consists of external and internal observations. External observations are conducted every 2 weeks (or possibly monthly), while the internal symptoms are observed at the end of each experiment. The appearance of *Ganoderma* mycelium, sporophore and foliar symptoms in seedlings both alive and dead is counted as being infected by external observation. The seedlings are considered infected if, by internal observation, there are damaged tissues inside the bole and/or trunk.

7.1   External Observations

External observations follow the method described by Breton *et al.* (2006) and Idris *et al.* (2006). Foliar discoloration, indicating the appearance of fungus symptoms, is counted as an infection. Fungus symptoms can also appear as mycelium and/or basidiocarp (fruiting body). Standardization of symptom scoring is important in an effort to develop high-volume screening.

Methods

1. Observe each tested seedling in the experiment.
2. At the first observation (4 weeks after planting), identify the seedlings that have failed to emerge and/or are abnormal. These seedlings should be replaced by normal seedlings.
3. Disease symptom observation starts at the second observation date (6 or 8 weeks after planting). Identify each seedling based on its specific label (Fig. 7.2).
4. All the symptoms are recorded on the observation form.

© Miranti Rahmaningsih, Ike Virdiana, Syamsul Bahri, Yassier Anwar,
Brian P. Forster and Frédéric Breton 2018. *Nursery Screening for Ganoderma*
Response in Oil Palm Seedlings: A Manual   55

Table 7.1. Materials, equipment and tools needed for external observations.

Materials	Equipment and tools
Oil palm seedlings inoculated by *Ganoderma*	Observation form (Fig. 7.1) Pen

Expt No:　　　　　　　　　　　　　　　　Recorder:
Date:　　　　　　　　　　　　　　　　　　Estate/Div/Block:

Plot	Palm No	Rep	Infected by *Ganoderma*	Died by *Ganoderma*	Basidiocarp Appearance (Yes/No)					
					FB	M	LS	MS	FBS	FBP
1	1									
	2									
	3									
	4									
	5									
	6									
	7									
	8									
	9									
	10									

Fig. 7.1. Example of an observation form for *Ganoderma* seedling screening at a nursery.

7.2　Internal Observations

Internal observations are conducted with a destructive treatment of the seedlings. The seedling is split by making a longitudinal cut in the bole. The severity of internal symptoms is assessed by visual estimation of the amount of damaged tissues caused by *Ganoderma* according to a scale established by Breton *et al.* (2006) (Table 7.2).

Methods

1. Pull out each seedling and split (using a knife) by making a longitudinal cut in the bole. Identify each seedling based on the specification.
2. All the symptoms are recorded on the observation form.

Fig. 7.2. External observation specification consists of healthy seedling (a) and infected seedlings (b, c, d) with dry leaves (arrow in b), *Ganoderma* mycelium and dry leaves (arrows in c) and *Ganoderma* basidiocarp and dry leaves (arrows in d).

Table 7.2. Disease severity scale based on internal observation. (From Breton *et al.*, 2006.)

Scale	Explanations
0	Healthy: no internal rot
1	Up to 20% rotting of bole tissues
2	From 20% to 50% internal rotting
3	Over 50% internal rotting
4	Total rotting of bole tissues along with total desiccation of the plant

Fig. 7.3. Internal observation scale: (a) healthy seedling and (b) infected seedling with 100% (total) rotting of bole tissues.

Table 7.3. Materials, equipment and tools needed for internal observation.

Materials	Equipment and tools
Oil palm seedlings inoculated by *Ganoderma*	Knife Observation form (Fig. 7.1) Pen

References

Breton, F., Hasan, Y., Hariadi, Lubis, Z. and De Franqueville, H. (2006) Characterization of parameters for the development of an early screening test for basal stem rot tolerance in oil palm progenies. *Journal of Oil Palm Research* Special Issue April, 24–36.

Idris, A.S., Kushairi, D., Ariffin, D. and Basri, M.W. (2006) Technique for inoculation of oil palm germinated seeds with *Ganoderma*. *MPOB Information Series* 314, 321–324.

Future Possibilities 8

Abstract

It has been claimed that *Ganoderma*-resistant oil palm material holds the greatest hope for the future control of basal stem rot (BSR) in South-east Asia. Conventional breeding currently requires germplasm with resistance and a phenotypic screening in both the nursery and the field. A better method might be provided by genotypic screening, which is independent of the environment and which can be exploited by breeders using marker-assisted selection. The identification of genes and the development of DNA diagnostics for *Ganoderma* resistance would be a major step forward in developing disease-resistant oil palm. However, to develop this, it is first necessary to identify resistant genotypes.

8.1 Gene or Marker Detection for *Ganoderma* Partial Resistance

The use of marker-assisted selection (MAS) using DNA markers is a useful tool in accelerating plant breeding (Forster *et al.*, 2015). Both quantitative and qualitative traits can be analysed and selected using these techniques.

One gene that might have a role in plant disease resistance is *EgNOA1* (GenBank accession number KF601427). Real-time PCR (qPCR) analysis revealed that the transcript abundance of *EgNOA1* in root tissues was increased by *Ganoderma boninense* treatment. This induction is an indication of the oil palm defence system activation against invading pathogenic *G. boninense* (Kwan *et al.*, 2014). Moreover, based on the transcript level in gene expression, there are six housekeeping genes (controlling β-actin, cyclophilin, GAPDH, MSD, NAD and ubiquitin), whose expression remains stable in oil palm root tissues after fungal infection, and which are recommended as reference genes (Kwan *et al.*, 2016).

It was reported in 2011 that Felda Agricultural Services Sdn Bhd, an oil palm research station in Malaysia, had created the world's first marker to detect *Ganoderma* disease (Publication No WO 2013066144 A1; Zulzaha,

© Miranti Rahmaningsih, Ike Virdiana, Syamsul Bahri, Yassier Anwar, Brian P. Forster and Frédéric Breton 2018. *Nursery Screening for Ganoderma Response in Oil Palm Seedlings: A Manual*

2011). Although the marker has not been able to identify the resistant material directly, it is expected to assist in screening pedigrees and then to produce *Ganoderma* partial resistant materials. Five years after the creation of the marker, Felda produced the first *Ganoderma* partial resistant oil palm materials in Malaysia (Anon., 2016). Other oil palm seed companies have released resistant materials using this genomics method to identify resistant materials.

Besides using gene markers, alteration in the protein abundance of infected oil palms may be used to identify the resistance or susceptibility of oil palm materials. In this method, changes are detected in the expression of proteins (proteomics) during infection to disease resistance processes. This has been utilized in other crops such as rice (rice blast fungus) and wheat (*Fusarium* disease) (Al-Obaidi *et al.*, 2014).

8.2   *Trichoderma* and Other Biocontrol Agents

The use of *Trichoderma* as a biological control agent for BSR disease is effective in preventing disease spread (see Chapter 1 of this manual). *Trichoderma* suppresses *Ganoderma* through competition for space and nutrition (Susanto *et al.*, 2013). *Trichoderma* isolates with the potential to antagonize *Ganoderma* are selected in a dual-culture assessment system. This is a simple method to gain information on the aggressiveness of *Trichoderma* against *Ganoderma* before being tested in nursery or field conditions (Sundram and Idris, 2009). A piece of *Ganoderma* mycelium is placed on a medium opposite a test isolate of *Trichoderma* (Fig. 8.1.a). The antagonistic potential of *Trichoderma* isolates is assessed after 5–8 days by measuring the radial growth of *Ganoderma* in the direction towards the growth of the *Trichoderma* isolate (Fig. 8.1.b) (Sundram, 2013).

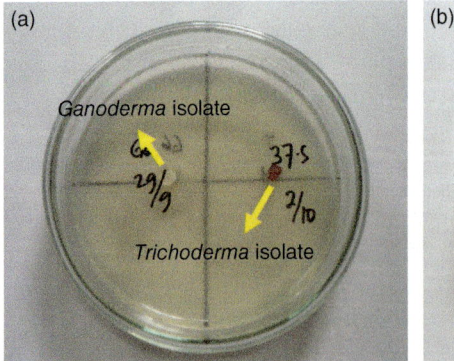

Fig. 8.1. Dual-culture assay between *Trichoderma* and *Ganoderma*: (a) a piece of *Trichoderma* and a piece of *Ganoderma* are planted on a potato dextrose agar (PDA) growing medium (arrowed); (b) an aggressive *Trichoderma* isolate (right) showing dominance over *Ganoderma* (left).

The effectiveness of *Trichoderma* is validated through root artificial inoculation in the nursery (Fig. 8.2) (Izzati and Abdullah, 2008; Sundram and Idris, 2009). Rubber wood blocks inoculated by *Ganoderma* provide an inoculum source in the nursery polybag (Chapters 5 and 6 of this manual). The antagonism of *Trichoderma* to *Ganoderma* is assessed by the percentage of seedlings showing *Ganoderma* symptoms or as a disease severity index (DSI). *Trichoderma* can reduce *Ganoderma* infection significantly in the nursery, particularly when the antagonist is in direct contact with the pathogen (Fig. 8.3) (Virdiana *et al.*, 2012). This method is being used to screen for the most potent *Trichoderma* isolate before deployment in the

Fig. 8.2. (a) Nursery screening of *Trichoderma*; (b) seedling infected by *Ganoderma*.

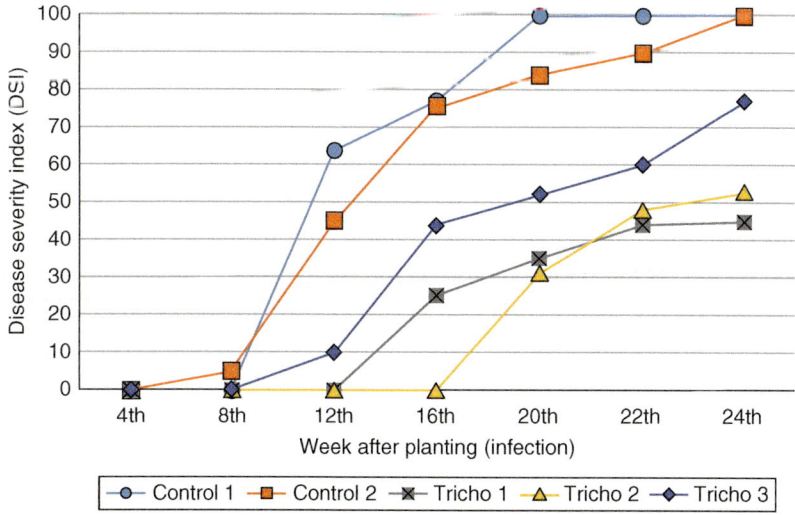

Fig. 8.3. Disease progression of BSR symptoms throughout the 24 weeks of observation in a seedling nursery test. (From Sundram and Idris, 2009.)

field as a biological control agent (Sundram and Idris, 2009). Currently, some plantation companies apply *Trichoderma* to polybags in the nursery to protect palm seedlings.

Alternatively, some species of bacteria, such as *Burkholderia cepacia* and *Pseudomonas aeruginosa*, are also able to inhibit the growth of *G. boninense* in oil palm plantations. It is reported that inoculation of *P. aeruginosa* reduced BSR incidence in seedling trials by 76%, while *B. cepacia* reduced incidence by 42%, and the mixture of both species by 54% (Sapak *et al.*, 2008). In addition, soil health plays an important role in oil palm plantations, because the ready existence and high abundance of beneficial microbes can supress *Ganoderma*.

It is essential to observe the microbe–plant interaction during disease infection and remedial action taken as early as possible; this can be done by testing at the DNA level. The ability of oil palms to perceive molecular signals and their response from beneficial organisms or disease-causing agents is critical to determine the susceptibility or defence response from oil palms towards harmful and beneficial organisms.

8.3   Other Oil Palm Seedling Screening Targets

Stress in crops is not only due to a pest or a disease; other environmental stresses also affect the growth and yield of oil palm. These include water deficit or drought, nutrient deficiency, high temperature and other environmental stresses. Environmental stresses like drought affect the growth of the palm, the ratio of male:female bunches, the rate of aborted female inflorescences and, eventually, yield (Rival, 2017). Previously, in order to test any environmental effects on oil palm yield, multi-genotype × multi-environment trials (GxE) were conducted. These trials aim to identify the best-adapted genotypes in relation to resource use efficiency (light, water, nutrients) and stress adaptation traits (Soh, 2016). However, GxE trials require large land areas and, in perennial species like oil palm, a long time. Another approach is to develop pre-field screens; for example, by conducting seedling tests of abiotic stress. The methods described above for *Ganoderma* seedling screening may be adapted for these purposes. In addition, physiological and genetic studies are needed urgently to understand and exploit responses to abiotic stress in oil palm. Since the sequences of the genomes of both oil palm species (*Elaeis guineensis* and *Elaeis oleifera*) have been made public, the genetic exploration of genes controlling resistance to biotic and abiotic stress has been thrown open (Singh *et al.*, 2013). Currently, the Malaysian Palm Oil Board (MPOB) has released molecular tools to identify shell thickness, fruit colour and mantled in oil palm (Singh *et al.*, 2014, 2015; Ong-Abdullah *et al.*, 2016). This marks the beginnings of MAS in oil palm, and it is expected to expand to biotic and abiotic resistance in the future.

References

Al-Obaidi, J.R., Mohd-Yusuf, Y., Razali, N., Jayapalan, J.J., Tey, C.C. *et al.* (2014) Identification of proteins of altered abundance in oil palm infected with *Ganoderma boninense*. *International Journal of Molecular Sciences* 15(3), 5175–5192.

Anon. (2016) Media release: FGV launch Malaysia's first *Ganoderma* tolerant oil palm planting material. Felda Press Centre. Available at: http://www.feldaglobal.com/media-release-fgv-launch-malaysias-first-ganoderma-tolerant-oil-palm-planting-material/ (accessed 26 March 2018).

Forster, B.P., Till, B.J., Ghanim, A.M.A., Huynh, H.O.A., Burstmayr, H. and Caligari, P.D.S. (2015) Accelerated plant breeding. *CAB Reviews No* 43, pp. 1–16. Available at: https://www.cabi.org/cabreviews/review?review=20153213434 (accessed 10 April 2018).

Izzati, M.Z.N.A. and Abdullah, F. (2008) Disease suppression in *Ganoderma*-infected oil palm seedlings treated with *Trichoderma harzianum*. *Plant Protection Science* 44(3), 101–107.

Kwan, Y.M., Meon, S., Ho, C.L. and Wong, M.Y. (2014) Cloning of nitric oxide associated 1 (NOA1) transcript from oil palm (*Elaeis guineensis*) and its expression during *Ganoderma* infection. *Journal of Plant Physiology* 174, 131–136.

Kwan, Y.M., Meon, S., Ho, C.L. and Wong, M.Y. (2016) Selection of reference genes for quantitative real-time PCR normalization in *Ganoderma*-infected oil palm (*Elaeis guineensis*) seedlings. *Australasian Plant Pathology* 45(3), 261–268.

Ong-Abdullah, M., Ordway, J.M., Jiang, N., Ooi, S.E., Mokri, A. *et al.* (2016) SureSawit™ Karma – a diagnostic assay for clonal conformity. *MPOB TS No.* 156 (June), 738.

Rival, A. (2017) Breeding the oil palm (*Elaeis guineensis* Jacq.) for climate change. *OCL* 24(1), D107. Available at: https://doi.org/10.1051/ocl/2017001 (accessed 10 April 2018).

Sapak, Z., Meon, S. and Ahmad, Z.A.M. (2008) Effects of endophytic bacteria on growth and suppression of *Ganoderma* infection in oil palm. *International Journal of Agriculture and Biology* 10, 127–132.

Singh, R., Ong-Abdullah, M., Low, E.T.L., Manaf, M.A.A., Rosli, R. *et al.* (2013) Oil palm genome sequence reveals divergence of interfertile species in old and new worlds. *Nature* 500, 335–341.

Singh, R., Ti, L.L.E., Ooi, C.L.L., Ong-Abdullah, M., Manaf, M.A.A. *et al.* (2014) SureSawit™ Shell – a diagnostic assay to predict oil palm fruit forms. *MPOB TT No* 548 (June), 656.

Singh, R., Ooi, L.C.C., Low, L.E.T., Ong-Abdullah, M., Nagappan, J. *et al.* (2015) SureSawit™ Vir – a diagnostic assay to predict colour of oil palm fruits. *MPOB TT No* 568 (June), 697.

Soh, A.C. (2016) Breeding for climate change mitigation and adaptation in oil palm. In: Rajanaidu, N. and Jalani, B.S. (eds) *Proceeding of the 5th International Conference on Oil Palm and Environment (ICOPE)*, 16–18 March 2016, Bali, Indonesia, CIRAD.

Sundram, S. (2013) First report: Isolation of endophytic *Trichoderma* from oil palm (*Elaeis guineensis* Jacq.) and their *in vitro* antagonistic assessment on *Ganoderma boninense*. *Journal of Oil Palm Research* 25(3), 368–372.

Sundram, S. and Idris, A.S. (2009) *Trichoderma* as a biocontrol agent against *Ganoderma* in oil palm. *MPOB Information Series TT* 422 June 2009. ISSN 1511-7871.

Susanto, A., Prasetyo, A.E. and Wening, S. (2013) Infection rate of *Ganoderma* at four soil texture classes. *Jurnal Fitopatologi Indonesia* 9(2), 39–46.

Virdiana, I., Anjara, P., Flood, J., Sitepu, B., Hasan, Y. *et al.* (2012) Replanting system and *Trichoderma* as an effective form of *Ganoderma* control and proven results. *4th Oil Palm Summit*, 9–10 July, Denpasar, pp. 1–10.

Zulzaha, F.F. (2011) Felda creates world's first marker to detect *Ganoderma* disease. *The Star Online*. Available at: http://www.thestar.com.my/business/business-news/2011/12/16/felda-creates-worlds-first-marker-to-detect-ganoderma-disease/ (accessed 26 March 2018).

Index

Page numbers in **bold** type refer to figures and tables.

CABI – who we are and what we do

This book is published by **CABI**, an international not-for-profit organisation that improves people's lives worldwide by providing information and applying scientific expertise to solve problems in agriculture and the environment.

CABI is also a global publisher producing key scientific publications, including world renowned databases, as well as compendia, books, ebooks and full text electronic resources. We publish content in a wide range of subject areas including: agriculture and crop science / animal and veterinary sciences / ecology and conservation / environmental science / horticulture and plant sciences / human health, food science and nutrition / international development / leisure and tourism.

The profits from CABI's publishing activities enable us to work with farming communities around the world, supporting them as they battle with poor soil, invasive species and pests and diseases, to improve their livelihoods and help provide food for an ever growing population.

CABI is an international intergovernmental organisation, and we gratefully acknowledge the core financial support from our member countries (and lead agencies) including:

  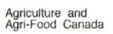  

Discover more

To read more about CABI's work, please visit: **www.cabi.org**

Browse our books at: **www.cabi.org/bookshop**,
or explore our online products at: **www.cabi.org/publishing-products**

Interested in writing for CABI? Find our author guidelines here:
www.cabi.org/publishing-products/information-for-authors/